Transient Airflow in Building Drainage Systems

T0173989

Transient Airflow in Building Drainage Systems

John Swaffield

Spon Press
an imprint of Taylor & Francis

LONDON AND NEW YORK

First published 2010 by Spon Press
2 Park Square, Milton Park, Abingdon, Oxon OX14 4RN

Simultaneously published in the USA and Canada by Spon Press
52 Vanderbilt Avenue, New York, NY 10017

First issued in paperback 2020

Spon Press is an imprint of the Taylor & Francis Group, an informa business

© 2010 John Swaffield

Typeset in Times New Roman by
HWA Text and Data Management, London

This publication presents material of a broad scope and applicability.
Despite stringent efforts by all concerned in the publishing process, some
typographical or editorial errors may occur, and readers are encouraged to
bring these to our attention where they represent errors of substance. The
publisher and author disclaim any liability, in whole or in part, arising from
information contained in this publication. The reader is urged to consult with
an appropriate licensed professional prior to taking any action or making any
interpretation that is within the realm of a licensed professional practice.

British Library Cataloguing in Publication Data
A catalogue record for this book is available from the British Library

Library of Congress Cataloging-in-Publication Data
Swaffield, J. A., 1943–
  Transient airflow in building drainage systems / John Swaffield.
      p. cm.
  Includes bibliographical references and index.
  1. Plumbing—Waste-pipes—Simulation methods. 2. Drainage—
  Simulation methods. 3. Ventilation—Simulation methods. 4. Air flow—
  Simulation methods. I. Title.
  TH6658.S93 2010
  696'.1—dc22   2009040732

ISBN 13: 978-0-367-57718-6 (pbk)
ISBN 13: 978-0-415-49265-2 (hbk)

In memory of our daughter
Toni Elizabeth Swaffield
1969 to 2006

Contents

3 Mathematical basis for the simulation of low amplitude air pressure transients in vent systems

4 Simulation of the basic mechanisms of low amplitude air pressure transient propagation – AIRNET applications

Figures

Tables

Preface

The air pressure regime and entrained transient airflows established within building drainage and vent systems result from the random discharge of appliances throughout the network, system surcharge or the arrival of pressure transients propagated by events external to the building system itself, and belong to a clearly defined family of unsteady flow phenomena variously described as pressure surge or transient propagation, a grouping that also, historically, includes waterhammer. The air pressure regime is a direct result of the superposition of pressure transients generated as a result of a multiplicity of changes in the system operating condition and wave reflections and transmissions, while the resulting entrained airflow depends fundamentally upon the shear force between the applied water flow and the air within, or drawn into, the network.

Pressure surge has been the subject of an international research effort over the last century, since Joukowsky's seminal work at the St Petersburg Imperial Water Works in 1900. This research effort involved contributors from across the international community and led to an industry standard analysis and simulation capability based on the solution of the defining St Venant equations by means of the Method of Characteristics linked to finite difference techniques. The range of fluid networks and operations successfully and, more importantly, routinely, addressed includes major civil engineering projects such as hydroelectric power stations; water, oil or gas distribution networks; a wide range of cooling water networks, including nuclear, and more specialist applications such as in-flight refuelling and fire-fighting network provision.

The processes generating unsteady flow in building drainage and vent systems are analogous in all respects to these pressure surge examples, with the exception of scale – failure conditions in the traditional manifestations of pressure transient propagation will be measured in atmospheres; in building drainage the equivalent 'pressure surge' will be measured in mm of water gauge. However the potential for catastrophic failure remains – the fatalities directly attributable to the drainage network transient driven SARS virus spread through dry appliance trap seals during the Amoy Gardens 2003 event highlights the need to fully understand the unsteady flow phenomena that accompany appliance discharge to the drainage networks in complex buildings.

This book will emphasise the development of analysis and simulation techniques that draw upon pressure surge theory to deliver a methodology capable of modelling all of the transient propagation events likely within the operation of a building drainage and vent system. The fundamental mathematical methods will be presented and shown to apply to building drainage by reference to a research programme at Heriot Watt University that has, since 1989, developed a simulation model, AIRNET, through a major series of Research Council and industry funded research projects. This research included the initial development of the AIRNET model; extensive site testing to build a database of system pressures in response to applied flows; the development from this database of the fundamental shear force relationships that define entrained airflows; the development and incorporation of a database of system boundary conditions, compatible with an MoC representation of network operation, into AIRNET; application of the methodology to a comprehensive model of underground station drainage; application of the methodology to a forensic evaluation of the SARS virus transmission route through the drainage network at Amoy Gardens; the invention and development of positive transient attenuation, leading to the P.A.P.A.™ unit; introduction and application of the concept of Active Control of transients and a sealed building design methodology, and the invention of a non-invasive, non-destructive transient propagation based methodology to identify dry traps – a concept clearly influenced by the SARS epidemic.

Thus there is now available a comprehensive simulation methodology that, for the first time, provides the system designer with the means to predict the likely pressure regime and entrained airflow conditions that will follow from design decisions. This will allow a re-evaluation of the codified design guidance currently available featuring the cultural preferences of the particular national, or transnational, code committee responsible for the design guidance provided.

This book will therefore present the underlying theory behind transient analysis; demonstrate its application to a wide range of known system operating conditions; utilise the methodology to model installed system operation and to forensically analyse failure conditions; introduce new design approaches and system surge protection devices as well as detailing the defective trap seal identification methodology.

The consequences of this fundamental re-analysis of system operation will be compared with current code recommendations through a review of current international codes. Many contributed to this review, in particular Steve White (Studor), Peter White (Hoare Lea), Steve Cummings (Caroma) and Karel de Cuyper, CIBW62 Coordinator, who shared his memories of being a member of a European code committee, are to be thanked. Special thanks are due to Dr David Kelly who undertook the task of compiling the code comparisons for an imaginary building.

This book will therefore emphasise that drainage system design stands at a crossroads – it can continue to be based on codified design guidance, that in many cases cannot be fully supported through engineering analysis, or it can develop based on a new approach to the analysis of transient fluid flow conditions, thereby providing the designer with the ability to fully understand and react to the operational requirements of the system and the new system requirements that will inevitably arise as a result of water conservation and climate change, increased population concentrations in the mega cities of the developing world, the introduction of new appliances and design requirements and the need to revisit our perceptions of cross-contamination and infection control in the aftermath of the SARS epidemic. The text concludes that a re-evaluation of design guidance is necessary based on the laws of physics, inherent to the analysis and simulations presented, rather than a continuation of a reliance on rules of thumb and code accretion.

This book, and its underlying research base, would not have been possible without the support and commitment of many individuals and organisations, hopefully all recognised in the Acknowledgements.

University research, while retaining its academic objectives, must be aware of, and respond, to industrial and societal needs. In parallel, industry has to recognise that academia has a contribution to make. Many organisations have supported our research, however the Drainage Research Group at HWU has been particularly fortunate to have formed such an enduring partnership with Studor Ltd who had the imagination to recognise that university research can inform product and system design development. Work in conjunction with Studor led to the P.A.P.A.™ device, to the Active Control and sealed building concepts – as demonstrated by the O_2 Dome installation – and currently to the development of the defective trap seal identification methodology and equipment. Sture and Doris Ericson, Franz Arnold and Steve White and their colleagues have all contributed to the work presented in this book.

This book has drawn heavily upon the work of the Drainage Research Group at Heriot Watt University, research that encompasses four main strands, namely drainage system air pressure regime and air entrainment prediction, solid transport in attenuating water flows, siphonic roof drainage and water conservation and climate change. It has been a privilege and a pleasure to have worked with all the members of this team over so many years. Writing this text has brought back too many happy memories to mention, however it will become obvious to the reader that this book could not have been written without the body of research in this 'air' theme alone by Dr David Campbell, Dr Ian McDougall, Dr Lynne Jack, Dr Grant Wright, Dr Michael Gormley, Dr David Kelly, Dr Kenny MacLeod, Dr Steven Filsell, Richard Beattie, Charles Hartley, Dr Caben Thancanamootoo and Dr Larry Galowin, long time friend, co-worker and Leverhulme Visiting Professor 1998–99 at HWU. The support of David Walker, who responded to all our requests for technical assistance, in the laboratory or on some site, with such good humour was much valued, while Dr Scott Arthur and Dr Eddie Owens contributed their own research expertise in support of this theme.

At Taylor & Francis, Tony Moore, whose initial invitation resulted in this text, and Simon Bates, who has dealt with all my queries with patience, are to be especially thanked for bringing the project to a satisfactory conclusion. As always any errors of understanding are mine alone.

The book makes the argument for change in drainage and vent system design. It will be the responsibility of industry to take up this challenge and the responsibility of the international research community to provide the necessary underpinning that will lead to new, improved analysis and simulation techniques in support of a better understanding of drainage system design, an understanding that will banish the 'rule of thumb' – a new departure made imperative by the demands on building services design emanating from global climate change, population movement, economics and public health requirements.

John Swaffield
Old Cambus
Berwickshire
September 2009

Acknowledgements

This book has drawn upon the work of the Drainage Research Group at Heriot Watt University from 1985 to 2009. The group's research has concentrated on four themes within building drainage, namely solid transport in attenuating flows, vent system design, siphonic rainwater design and analysis, water conservation and climate change. This book concentrates on the air pressure and entrained airflow simulations developed within the ventilation system design theme as a continuous activity over this period.

This research theme has been supported by the UK Engineering and Physical Sciences Research Council and its predecessors throughout the period, a total of 19 person years of Research Associate support; that support is gratefully acknowledged.

The group has always believed that academic research should be linked to 'real' industry problems and this has been a central feature of this theme. Support from UK and international industry, professional institutions, code bodies and government over this period contributed a further 10 person years of Research Associate effort. The support of ASPE, Caroma, CIBSE, Defra, Hepworths (now Wavin UK), IAPMO, LUL, Marley, RAE, RSE, SOPHE, SNIPEF, Studor and WPC among others is gratefully acknowledged. In addition to financial support, the research group has always valued the guidance offered through project steering groups by consultants such as BuroHappold, Arup and Hoare Lea, commercial organisations such as RBS, Consort Healthcare, Balfour Beatty Workplace, IAPMO, SNIPEF and the UK Department of Health, among others.

The research community in building drainage and water supply for buildings has consistently, over 35 years, encouraged the work of the group at Heriot Watt and its predecessor group at Brunel University. The annual conferences of the CIBW62 Working Commission has provided a continuing source of support and a forum for the exchange of ideas with colleagues drawn from across the international research community. The close working links established have been a consistent source of support, historically with NBS Washington and more recently with National Taiwan University of Science and Technology, among others.

Heriot Watt University, its Department of Building Engineering and Surveying and more recently its School of the Built Environment, is to be thanked for the continuous support provided in the form of well-found laboratory and technical support, computing facilities and expert IT assistance whenever required.

Jeff Patchell, publisher of the journal *International Plumbing Review*, is to be thanked for making his journal available to the BBC satirical show 'Have I Got News For You' during its 2008 season, leading to the work of the HWU Drainage Research Group being featured in the 'missing words' round – a pleasant counterpoint to being an internationally leading research group.

Abbreviations and notation

Variables

A	Pipe cross-sectional area, m^2
$A_{1 \text{ to } n}$	Pipe cross-sectional areas at junction, m^2
A_{Trap}	Cross-sectional area trap, m^2
ACF	Absolute compliance factor
B_{1-4}	Coefficients defining appliance discharge under pressure
C	Chezy resistance coefficient
C	Ratio of entrained airflow to applied annular water flow
C^+, C^-	Characteristic equations sloping downstream and upstream respectively
C_R	Reflection coefficient
C_T	Transmission coefficient
C_j	Error in estimation of trap position due to junction effect
C_{0-3}'	Fan characteristic coefficients
c	Wave speed, m/s
c_0	Wave speed in unconfined fluid, m/s
D	Duct diameter, m; fan diameter, m
D_a	Entrained air core diameter in vertical stack, m
D_p	Simulated Surcharge pressure rise, mm
DU	Appliance discharge unit
dH	Water seal differential, mm
E	Young's modulus of elasticity, N/m^2
e	Pipe wall thickness, m
FU	Appliance fixture unit
F(t+x/c)	Wave propagating upstream
F(t–x/c)	Wave propagating downstream
$F_{1 \text{ to } n}$	Transients propagated upstream within network
F_c	Circumferential stress, N/m^2
$f_{1 \text{ to } n}$	Transients propagated downstream within system
f	Friction factor, Colebrook–White or empirical after Jack (2000)
f_{wet}	Empirical friction factor providing pressure recovery in a wet stack

g	Acceleration due to gravity, m/s^2
H_1	Trap seal water level, system side, mm
H_0	Trap seal water level, appliance side, mm
H_w	Wet stack height, m
J_ε	Equivalent pipe length attributed to the presence of junctions in trap seal identification study
K	Bulk modulus fluid, N/m^2, separation loss at entry, junction or exit to the stack
K	Constant in gas equation; constant in terminal velocity expression
K_c	Vent exit loss during closure
K_{eff}	Effective bulk modulus inc. gas and wall effect, N/m^2
K_{gas}	Bulk modulus gas, N/m^2
K_{Max}	Maximum exit loss representing vent closure
K_0	Vent exit loss during opening
K1, K2, K3, K4	Constants in the finite difference form of the St Venant equations
K_1	Constant in expression linking entrained airflow to stack diameter and annular flow area ratio
K_2	Constant in expression linking annular water downflow to stack diameter and annular flow area ratio
K_0	Fully open loss for a duct damper
$K_{0.5}$	50% closed loss for a duct damper
k	Pipe wall roughness, mm, constant in gas equation
L	Pipelength, m; length of trap seal water column, m
L_1, L_2	St Venant equations of continuity and momentum
L_p	Line pressure, N/m^2
MD	Maximum deviation, defect free, cf. test pressure
m	Mass, kg, and velocity profile coefficient in unsteady friction treatment, hydraulic mean depth, m
m_w	Mass of water in trap, kg
mSp_{DF}	Mean defect-free pressure at any location
mSp_{TEST}	Test pressure at any location
N	Number of internodal reaches in any pipelength; number of defect-free baseline tests
N	Fan speed, rads/s
N	Velocity profile coefficient in unsteady friction treatment
$N+1$	Number of nodes in any pipelength
n	Manning roughness coefficient, number of appliances of a particular type
n	Number of junctions traversed by test transient in trap seal identification study
P	Pipe wetted perimeter, m
P	Probability that any number r of n appliances are in use at the same time
p	Pressure, N/m^2

$P_{atmospheric}$	Atmospheric pressure, N/m^2
P_{branch}	Pressure in system branch pipe, N/m^2
p_w	Change in air pressure induced by appliance discharge
p_0	Ambient pressure, N/m^2
Q	Air or water flow, m^3/s
Q	Fan throughflow, m^3/s
Q_A, Q_a	Airflow rate, m^3/s
Q_W	Water flow rate, m^3/s
q	Lateral inflow in general momentum equation, m^3/s
R	Ratio D_p/L_p
Re	Reynolds number
r	Number of appliances of a particular type in simultaneous use
r_s	Ratio of annular flow area to wet stack cross-sectional area
T	Interval between uses of a particular appliance
T	Temperature
T_C	Closing time for a vent, s
T_O	Opening time for a vent, s
t	Duration of discharge of an appliance
t	Terminal annular flow thickness, mm
t, t'	Time, s
t^*	Non-dimensional time in unsteady friction treatment
t_p	Pipe period (2L/c), s
u	Flow velocity, m/s
$u_{1,2}$	Variables in St Venant equations
u_t	Trap water velocity, m/s
V	Mean fluid velocity in general unsteady friction treatment, m/s
V	Vehicle velocity in moving boundary equation treatment, m/s
V_{AIR}	Entrained airflow velocity at water interface, m/s
V_a	Mean entrained airflow velocity, m/s
V_t, V_{tl}, V_{Term}	Annular water film terminal velocity, m/s
V_{Water}	Annular water velocity at air interface, m/s
Vol_{gas}	Gas volume in a pipe section, m^3
Vol_{pipe}	Volume of pipe section, m^3
Vol_{total}	Total volume, gas plus pipe section, m^3
W	Weighting factor, unsteady friction treatment
X_D^{adj}	Adjusted predicted trap location to allow for junction effect
X_D^*, X_D^{true}	Trap position, estimated from wave speed and true
x	Distance, measured positive upwards in water flow direction, m
xt, xl, xu	Mean, lower and upper x values in a bisection technique iterative solution
Z_t	Vertical fall to achieve terminal velocity, m
z	Height above stackbase, m
α	Pipe slope, open setting of a duct damper
ΔH	Trap seal deflection, m
ΔP	Individual pressure losses, regains, in vertical stack, N/m^2

ΔQ	Incremental change in flowrate at junctions, m^3/s
Δp	Change in pressure, N/m^2
Δp	Pressure rise across a fan, N/m^2
ΔT_C	Time into vent closure, s
ΔT_O	Time into vent opening, s
Δt	Time increment, s
Δu	Change in velocity, m/s
Δx	Pipe section length, m
δp	Incremental change in pressure, N/m^2
$\delta \rho$	Incremental change in density, kg/m^3
δs	Incremental pipe section, m
δu	Incremental change in flow velocity, m/s
δx	Incremental length, m
γ	Ratio of specific heats
θ	$\Delta t / \Delta x$, V/V_t
λ	Multiplicative factor in MoC
μ	Dynamic viscosity, kg/ms
ρ	Density, kg/m^3
τ_0	Wall shear stress, N/m^2
υ	Kinematic viscosity, m^2/s

Subscripts and superscripts, used singly or in combination as appropriate

A, B, C, R, S	Nodes with known values of p, c, u at time t
Atm	Atmospheric conditions
bag	P.A.P.A. volume definition
branch	Branch properties
Continuation	Pipe carrying the transmitted transient past a junction
Dry stack	Drainage stack not carrying an annular water flow
f	Fluid
fan	Fan conditions or interface
g	Gas
Incoming	Pipe carrying the transient approaching a junction
i	Pipe identifier
inner	Inner diameter, differential trap
j	Pipe identifier
Local	Local conditions
MAX	Maximum values
open	Open end conditions
outer	Outer diameter, differential trap
P	Node with unknown values of p, c, u at time $t + \Delta t$
P.A.P.A.	Positive Air Pressure Attenuator
pipe	Pipe conditions
piston	Conditions at the pistonface during motion

ref	Fan reference speed, pressure, flow throughflow
room	Room conditions
SEWER	Sewer conditions
stack	Stack conditions or properties
sum	Summated stack pressures
sump	Sump conditions
sumpair	Sump air volume, m^3
surface	Conditions at the water to air interface in annular stack flow
surge	Transient pressure
Tank	Holding tank properties
Tankair	Holding tank air volume
Term	Terminal annular flow conditions
t, trap	Trap conditions
$t, t + \Delta t$	Time
vent	Vent airflow in below ground drainage systems
Wet stack	Drainage stack carrying an annular water flow
w, water	Water flow
0	Initial conditions
$1, 2, 3, ..., n$	Pipe identification

Abbreviations

AAV	Air Admittance Valve
ACF	Absolute Compliance Factor
AIRNET	HWU developed transient air pressure simulation
ASPE	American Society of Plumbing Engineers
AUS/NZ	Australian and New Zealand Standards Organisation
BBA	British Board of Agrément
BRE	Building Research Establishment (UK)
BSI	British Standards Institute
CF	Compliance Factor
CFD	Computational Fluid Dynamics
CIBW62	International Council for Research and Innovation in Building and Construction Working Commission W62 'Water Supply and Drainage for Buildings'
EPSRC	Engineering and Physical Sciences Research Council
FM4AIR	MoC simulation for duct work airflows
HEPA	High Efficiency Particulate Absorbtion filters
HSIDU	High Security Infectious Disease Units
HWU	Heriot Watt University
IAPMO	International Association of Plumbing and Mechanical Officials
IPC	International Code Council Inc.
LDA	Laser Doppler Anemometer
LUL	London Underground Ltd
LULVENT	MoC simulation developed for LUL applications

MoC	Method of Characteristics
MD	Maximum deviation
NBS	National Bureau of Standards, Washington DC, now NIST (National Institute for Standards and Technology)
NRV	Non-return valve
P.A.P.A.™	Positive Air Pressure Attenuator invented by Swaffield and Campbell, marketed by Studor Ltd
PTA	Positive Transient Attenuator*
PTG	Pressure Transient Generator
RAE	Royal Academy of Engineering
RBS	Royal Bank of Scotland
RSE	Royal Society of Edinburgh
SNIPEF	Scottish and Northern Ireland Plumbers Employers Federation
SoPHE	Society of Public Health Engineers
UPC	Uniform Plumbing Code
VHF	Viral Haemorrhagic Fever
VVCD	Variable Volume Containment Device, generic term to describe action of the P.A.P.A.™ unit
WHO	World Health Organisation
WPC	World Plumbing Council

The generic terms Variable Volume Containment Device (VVCD) and Positive Transient Attenuator (PTA) will be used in this book to describe the Active Control of positive transients by the introduction of a theoretical expandable containment volume. This generic approach was translated into a product invented by HWU and developed in conjunction with Studor Ltd who now market the product as the P.A.P.A.™. The trademark name will be used whenever the product is discussed.

1 Building drainage and vent systems, a traditional building service requiring an engineering analysis makeover?

The provision of reliable sanitation is a prerequisite of a developed society; this is as true for the mega-cities of the twenty-first century as it was for the emerging Greek civilisations around Homer's 'wine dark sea'. The pressures of population and climate change present new challenges to the design engineer that may not be met merely by reference to past experience or to codified guidance. Drainage and vent system design for the highly complex structures now being developed requires that the engineer has access to proven and fundamental analysis techniques that will, at the very least, provide the same level of confidence as that ascribed to other building services simulations.

In order to provide that understanding it is necessary to redefine the operation of building drainage and vent systems in terms of their fundamental operational characteristics and then draw upon the accumulated knowledge of fluid mechanics to provide analysis and simulation techniques that can contribute to improved system design.

While the perceived objective of a drainage system may be the removal of waste to the sewer, fundamentally an unsteady free surface liquid flow operation, the prevention of cross-contamination is an absolute necessity to protect against both infection spread and odour ingress. Therefore the analysis of building drainage and vent system operation must emphasise the mechanisms of air entrainment that accompany the unsteady liquid flows through the network and determine both the air pressure regime within the system and the probability of appliance trap seal depletion and cross-contamination. That this is an issue worthy of fundamental consideration was evidenced by the fatalities during the SARS epidemic in Hong Kong in 2003; an infection spread identified by WHO (2003) as exacerbated by poor drainage design and trap seal depletion.

The movement of entrained air within a building drainage and vent system is readily identified as a two-phase fluid flow phenomenon driven by the shear forces between the appliance water discharges and the air within the network, initially quiescent and at atmospheric pressure. The unsteady nature of the water flows inevitably result in an unsteady entrained airflow where the changes in airflow demand, as a result of the random discharge of the system appliances, are communicated through the propagation of low amplitude air pressure transients.

The pivotal role of transient propagation in establishing the air pressure regime within the network immediately links the analysis of building drainage and vent systems to the much wider and well-established field of pressure surge analysis. Developed over 100 years through a truly international research effort, pressure surge analysis is now an established component of fluid system design with industry standard analysis techniques based around the application of the mathematical Method of Characteristics and computer simulation. While failure conditions in pressure surge analysis may involve pressure surge predictions measured in atmospheres, the analogous failure conditions in drainage and vent system applications may only be 10s of mm of water gauge – sufficient to deplete essential appliance trap seals and lead to fatalities through cross-contamination. However the techniques employed to analyse pressure surge in a whole range of flow situations, from water supply or long distance oil or gas distribution to in-flight refuelling, are wholly transferable to the study of air pressure transient propagation in drainage networks; the fluid properties, the pipeline materials and the wave speed magnitudes may change but the defining St Venant equations of unsteady flow continuity and momentum remain identical. Many of the boundary conditions that will be met in drainage applications, including junction representation, relief valves and air chambers, mechanical interfaces to fans, pumps, variable setting valves, the reflections at system terminations and the treatment of unsteady fluid friction, all have direct analogues in pressure surge analysis. Similarly the more specialised applications of air pressure transient analysis to such areas as train and elevator motion within tunnels and shafts are directly related to the analysis techniques that will be developed.

This book will therefore take as its main objective the reassessment of the analysis of entrained air movement within building drainage and vent systems. The historical development of drainage and vent systems will be briefly presented together with a development of the essential relationships linking the water and airflows, including the methodology required to estimate system water throughflow as a result of random appliance discharge.

The fundamental concepts of pressure surge analysis will be presented together with a full development of the Method of Characteristics predictive and simulation techniques, drawn from the wider pressure surge research area. Examples of the application of these techniques will be presented to demonstrate the operational characteristics of a drainage network in the abstract. This will allow a better understanding of the overall conditions that determine the air pressure regime and the possibility of trap seal depletion. The theoretical aspects of transient analysis will be informed by drawing on a twenty-year research-led development of these analysis methods at Heriot Watt University where the Drainage Research Group has been continuously funded by government and international industry to develop these methodologies.

Applications to current building design processes, as well as forensic investigations aided by the analysis methods introduced, will be presented along with the theoretical base for a series of innovative products and system monitoring methodologies. The concept of Active Control to provide pressure surge control and suppression will be discussed and demonstrated, along with the basis for the sealed building design methodology.

The book will also identify areas where the national codes that have grown up over decades based on experiential knowledge, as opposed to the laws of physics that do not recognise national preferences, should take note of the predictive capability of the simulation techniques presented as a means of providing international standards.

Building drainage and vent system design and analysis may be recognised as a branch of unsteady fluid mechanics in every way amenable to the application of the predictive and simulation techniques already established within pressure surge analysis. This book aims to extend those methods to this undervalued branch of building services engineering.

1.1 The requirement for drainage and vent systems

The role of a building drainage system is to transport fluid, faecal and other waste from the appliance to the sewer connection with no possibility of cross-contamination, leakage or interruption to system availability. In addition, under the current conditions of climate change, the process should be undertaken with a minimum of water usage and the installed system should wherever possible use the minimum of resources.

The basic requirement for an efficient building drainage system, mirroring the provision of safe drinking water supplies, has been a prerequisite of civilisation over several thousands of years, from early and well-documented provision in the Eastern Mediterranean through the Roman era and the eventual reintroduction of sanitary knowledge to Europe, following the re-conquest of Spain, to the challenges posed by the development of the modern city, both as a result of the Industrial Revolution and now the worldwide migration of rural populations into the mega-cities of the developing world. While the development of modern sanitation is well documented, it will be helpful to highlight several key events as a basis for discussing the continuing design challenges to the successful provision of building drainage.

1.2 Basic operational mechanisms within drainage and vent systems

The operation of a building drainage and vent system involves the unsteady flow of water, where the time dependency depends upon the random operation of the appliances connected to the network, as well as the time dependent discharge profiles characterising each appliance, and the attenuation of any discharge wave during its passage through the network. In parallel a time dependent entrained airflow is established, where the time dependency arises as a result of the shear forces between the discharge water flow, which establishes an annular flow within the system vertical stacks, and the initially ambient stack air core. In addition, the transport of waste solids, ranging from faecal material to hygiene wipes in modern systems, is dependent upon the form of each appliance discharge and the position of the solids within the attenuating discharge wave. The first requirement is often seen as the successful transport of waste solids to the sewer connection without the normal periodic deposition, that must occur, leading to drain obstruction and interruption to service. Solid transport, also known as drainline carry, is therefore a prerequisite of successful drainage and vent system design and is covered by both codes and analysis techniques that recognise the importance of drain cross-sectional shape and dimension, as well as drain slope and appliance discharge volume. It is important to recognise that the flow conditions are time dependent so any steady state modelling of the drainage flow is flawed.

However, drainage design codes, linking applied water flowrates to drain diameter and slope, are based on the concept that the free surface flow depths should not exceed 50 per cent of the drain diameter and inherently imply steady flow. This result may be confirmed by application of the Chezy expression for steady free surface flow depth at the maximum allowable flow rate acceptable prior to either an increased diameter or a steepened slope recommendation. This essentially empirical result ensures that the increased flow depths generated at pipe junctions still allow an air path above the water free surface and do not result in local surcharge.

The time dependent waterflows within the drainage network entrain an airflow that is therefore itself unsteady. In order to communicate changes in shear driven airflow demand, each change in waterflow is accompanied by the propagation of a low amplitude air pressure transient that ensures that the modified airflow demand is recognised at the system boundaries. These pressure transients are capable of depleting the trap seals, which protect each appliance and prevent cross-contamination, as they propagate throughout the network. The appliance trap seal, originally invented by Alexander Cummings in 1775, is an essential element within any building drainage and vent system and simply consists of a U-tube filled with water that prevents cross-contamination of habitable space from the drainage network. Effectively a 'U-tube' manometer, this column of water responds to all the changes in pressure propagated as transients throughout the network. Trap seal depletion can lead to potentially disastrous consequences for the building users, as demonstrated later in the treatment of the 2003 SARS epidemic in Hong Kong. Similarly any sudden interruption to the airflow path, possibly due to local surcharge, results in the propagation of positive pressure transients that may well be sufficient to wholly deplete appliance trap seals.

Thus the flow conditions in a building drainage and vent system may be defined as unsteady multi-phase, multi-component flow. System design has historically been dependent on the development of rules of thumb. Modern numerical analysis methods, drawing upon experience in the allied study of pressure surge, currently offer the opportunity to improve the designers' understanding of the multiple flow conditions routinely met within these systems.

Figures 1.1 and 1.2 illustrate a simple drainage and vent network that may be used to illustrate the operational regime discussed above. Discharge of any one of the appliances within the system initiates a free surface flow in the connected

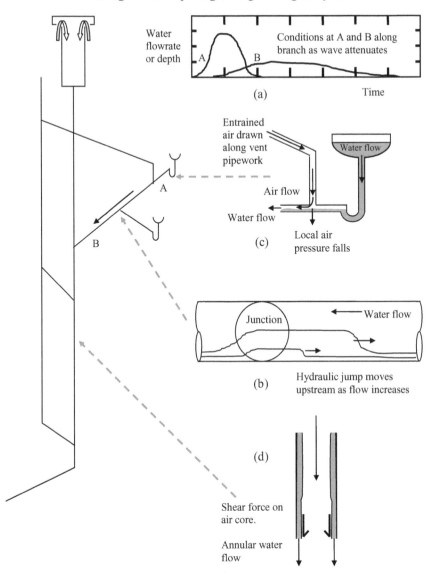

Figure 1.1 Component elements within a building drainage network, illustrating water flow conditions in both the system branches and vertical stacks

branch. This flow is time dependent at source as it reflects the discharge profile of the appliance. Washbasins, sinks and baths have a long duration discharge, as do washing machines and dishwashers, with a quasi-steady central portion to the discharge profile. However w.c. discharge is of a much shorter duration with a peak flow early in the profile. Such discharges attenuate along the branch as shown in Figure 1.1(a). Appliance discharges interact with the drainage network; at junctions

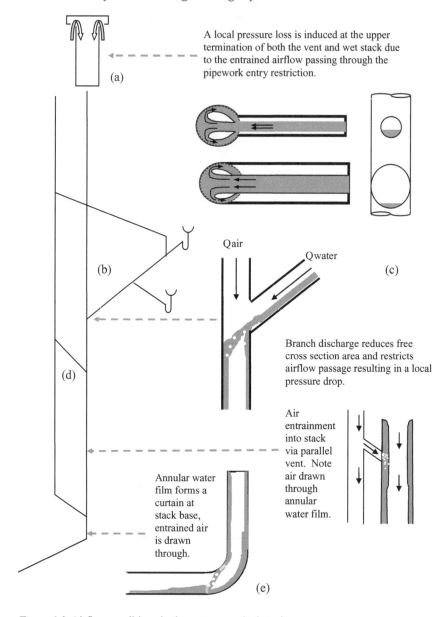

A local pressure loss is induced at the upper termination of both the vent and wet stack due to the entrained airflow passing through the pipework entry restriction.

(a)

(b)

(c)

Qair

Qwater

(d)

Branch discharge reduces free cross section area and restricts airflow passage resulting in a local pressure drop.

Air entrainment into stack via parallel vent. Note air drawn through annular water film.

Annular water film forms a curtain at stack base, entrained air is drawn through.

(e)

Figure 1.2 Airflow conditions in the system vertical stack

the joining flows are subject to the formation of a hydraulic jump in all the branches terminating at that junction. The position of the jump moves upstream, away from the junction, as the approach flowrate increases and falls back as it abates. This effect may lead to premature stranding of waste solids and to a possible surcharge condition where the entrained airflow is obstructed, Figure 1.1(b).

The discharge from the appliance also reduces the air pressure in the branch immediately downstream of the appliance trap seal, Figure 1.1(c). This reduction in local ambient pressure may cause trap seal loss at the end of the discharge and this is prevented by the addition of a local vent that provides an airflow into the branch and hence raises the local pressure, retaining the trap seal. This is only one of a number of trap seal failure conditions to be dealt with later. Trap seal depletion may also be prevented by careful design of the branch gradients and possibly increases in branch diameter.

As the appliance discharge reaches the system vertical stacks, which run the full height of the building collecting discharges from each floor, an annular flow regime is established, Figure 1.1(d). This annular water flow has an interface to the initially ambient pressure air core within the 'wet' stack and hence due to the action of the shear forces established at the fluid interface initiates a local air entrainment. This local airflow demand must be satisfied by an inflow of air to the network through its boundary terminations, whether through the open roof terminations or through a depleted trap seal or an in-building air admittance valve. In order for such airflows to be established, the demand has to be communicated to the system boundaries. This is achieved by the propagation of low amplitude air pressure transients that travel at the acoustic velocity in air and impose a pressure differential sufficient to generate the required airflow change – note that while increases in water flow generate transients that propagate a reduction in air pressure, a reduction in applied annular water flow would propagate a positive pressure change. Similarly any obstruction to the entrained airflow will propagate a positive pressure change.

The entrained airflow is subject to the established laws of flow resistance during its passage through the drainage network. As the local pressure at the roof line stack termination falls as the demand transient arrives, air is drawn into the upper stack, Figure 1.2(a), often referred to as the 'dry' stack, and therefore a local pressure drop is associated with this expected entry loss. Frictional losses then affect the airflow as it is established down the dry stack to the first discharging branch, Figure 1.2(b).

The appliance discharge enters the stack, again now often referred to as the 'wet stack', below the upper discharging branch, and forms a water curtain across the stack dependent upon the comparative diameters of branch and stack and the slope of the discharging branch, Figure 1.2(c). Airflow is drawn through this curtain and a local pressure loss is experienced.

Below the discharging branch the air core is subject to the interface shear forces already discussed and hence there is a pressure recovery zone that extends down to the stack base, or to any stack offset.

In drainage networks that feature a parallel vent stack, Figure 1.2, airflow is naturally established in the vent stack by reductions in pressure in the wet stack. This additional entrained airflow may enter the wet stack through the cross-connections illustrated in Figure 1.2(d). Again it is necessary for the entrained airflow to be drawn through the water curtain covering the cross-vent entry junction as shown.

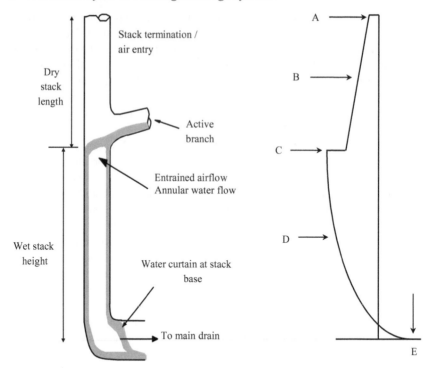

Figure 1.3 Basic mode of operation of a single stack building drainage network, illustrating the pressure regime established as a result of a single discharging branch

Finally the entrained airflow exits the stack via the stack base change of direction and connection to the horizontal drain connection to the sewer system. The transition from annular stack waterflow to free surface flow in the drain results in the formation of a further water curtain at the stack base through which the entrained airflow is forced by the pressure differential generated by the shear forces in the wet stack and the differential between this local terminal pressure and the drain pressure, normally taken to be close to atmospheric, Figure 1.2(e).

Figure 1.3 illustrates the pressure regime discussed above as established in a building drainage and vent system following the discharge of an appliance on one floor. A and C represent the local losses at entry and at the discharging branch, B and D represent the frictional resistance of the dry stack and the pressure recovery in the wet stack, while E represents the back pressure at the stack base.

Thus the pressure regime established within a building drainage and vent system may be discussed in terms readily understood in the general context of fluid mechanics, and the unsteady nature of both the water and entrained airflows may be recognised as examples of transient theory, a possibly more specialised branch of fluid mechanics, that offers access to established analysis and simulation techniques.

1.3   Historic development of building drainage and vent systems

While the drainage system has as a primary objective the removal of fluid and solid wastes from habitable space, the prevention of cross-contamination is also of primary concern. Initially, during the early Victorian development of building drainage and vent systems in the UK, the prevention of odour ingress to habitable space was a major concern, both from a user perspective and also as an influential body of opinion connected the spread of disease to odours – the miasma theory supported by Chadwick among others.

The earlier invention of the water trap seal in the late 1700s was seen as a solution that would allow a separation of the drainage system from the building interior. The simple introduction of a U-tube filled with water capable, due to a water column height designed to exceed any applied air pressure from the system, Figure 1.4, of preventing any odours within the system from penetrating into

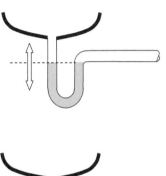

Under quiescent ambient conditions the trap seal appears as a U-tube with the water level controlled by the entry level to the appliance branch as shown. The trap seal depth is measured in mm of water gauge, 50 and 75 mm being commonly specified in national codes.

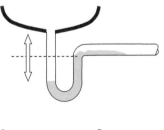

Negative pressure in the branch leads to a loss of trap seal water to the drainage network. Once the appliance side water column falls to the invert of the trap bend, a suction air path is formed and air is drawn into the network.

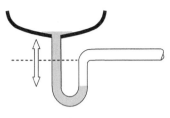

Positive pressure in the branch leads to a loss of trap seal water to the appliance. Once system side water column falls to the invert of the trap bend, an air path is formed and air is driven into habitable space via the appliance.

Appliance Side | System Side

Figure 1.4 Operation of an appliance trap seal to prevent cross-contamination of habitable space from the drainage network

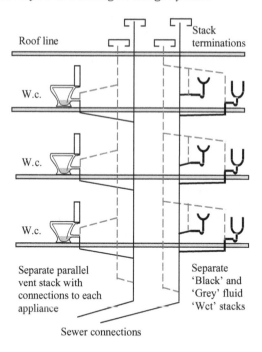

Figure 1.5 Original Victorian two-pipe system

habitable space, was a major advance that remains the first line of defence against cross-contamination. In parallel a complex venting network was also introduced, culminating in open upper terminations above the highest openable window in the building.

The Industrial Revolution, initially in the UK, led to two consequences, firstly mass migration to the developing city structures, leading to urban overcrowding and a real increase in disease, Chadwick (1842), and secondly, at the other end of the economic spectrum, the emergence of a sector of the population, the beneficiaries of the economic changes brought about by the Industrial Revolution, who demanded enhanced sanitation provision within their homes and other buildings. The invention of the appliance trap seal provided a methodology that contributed to both the general improvement of public health for the first group and the provision of acceptable sanitation for the second. Two hundred years later the basic design philosophy introduced at that time still dominates drainage and vent system design.

The earliest sanitation systems within this post-Industrial Revolution phase may be characterised by the so-called two-pipe system, Figure 1.5. In this solution each appliance is protected by a water trap seal that is in turn protected by a direct vent connection to a vent stack terminating above roof level. In addition the appliances are segregated into 'black' water discharges – w.c.s – and 'grey' water discharges, effectively all other appliances, with a separate wet stack provided for each, again terminating above roof level. Thus the two-pipe system featured four stacks as shown: two vents and two waste water stacks.

Obviously this design was both complex and expensive. Many examples do still remain visible, as the norm for the essentially low-rise UK applications during this period was to mount the network external to the building and many excellent examples may be seen on Victorian buildings – for example in Edinburgh where historic buildings built along the Royal Mile are effectively up to 10 storeys high due to their construction above and below the rising ridge line.

The latter half of the 19th century saw the first systematic investigation of building drainage and vent system operation and the identification of the underlying fluid mechanics. Gormley (2007b) identifies a series of ground-breaking contributions, many later forgotten or treated as too academic by the developing trade-biased plumbing profession. Gormley identifies the contribution in the UK of the eminent engineer Reynolds (1872), who contributed a paper on prevention of sewer gas contamination, and in the USA Waring (1895), sanitary engineering consultant to the President, who defined the operation of the vent system, correctly identifying for the first time the linkage between vent length and diameter for any particular pressure alleviation, clearly establishing drainage design as merely a branch of the developing subject of fluid mechanics and in particular identifying the central role of frictional representation.

At the turn of the century interest continued; in Boston Putnam (1911) queried the necessity for separate venting and effectively predicted the single stack system, now standard in the UK and the Philadelphia system, ASPE (2002), that is now gaining support in the USA, predating the central work of Wise (1952). Putnam recommended the use of 'mechanical vents' in close proximity to appliance trap seals, predating the air admittance valve and Active Control by 80 years. Gormley highlights Putnam's explanation for the failure of his forward concepts, namely the inherent conservatism of the plumbing associations then gaining ground in the USA. This is also related to the historic, and possibly continuing, divergence between 'plumbers' and 'sanitary engineers'. It is undeniable that drainage system design is not regarded as a 'prestige' role within building services engineering and this contributes to the development of the current position where national codes define the 'how' rather than the 'why' of drainage design, a possibly acceptable position until a new concept or threat requires a fundamental understanding of the fluid mechanics of the problem.

It is clear therefore that by the First World War there was a level of understanding of the operation of building drainage systems but that for a variety of reasons, related to labour issues, education and training, and the lack of the mathematical tools to deal with the inherently unsteady flows met in drainage networks, the practice of drainage design fell back on codified guidance rather than design and analysis. To some extent this continues to the present day. However, to put the lack of analytical methods in context, it is worth noting that while Putnam was predicting many of the subsequent design innovations of the mid-20th century, the analytical tools to deal with the pressure transient propagation that underpins all vent system operation was only just being developed by Joukowsky (1900) during his work on 'waterhammer' at the St Petersburg Imperial Water Works and the advent of fast computing, that transformed pressure surge analysis, was still 60 years in the future.

The traditional two-pipe system, developed during the late 1800s in the UK, was impractical for high-rise and internal positioning. Advances in high-rise construction along the US Eastern seaboard at the turn of the century led to the introduction of the one-pipe system, Figure 1.6, where the venting arrangement is retained but all appliances discharge to one stack and all appliance vents join a single vent stack, again both terminating above the roof line. This development was supported by some of the earliest recorded drainage research efforts, including an assessment of wet stack capacity, Hunter (1924), and later Babbit (1935) and Dawson and Kalinske (1937), work that is still regularly referenced in the US national codes. Cross-connection at the upper levels was also used to reduce the number of roof penetrations. By the 1930s the one-pipe system was in general use and remains so 80 years later, being the basis for much of the US design codes.

The shortages in building materials and the need to develop mass housing following the end of the Second World War, led researchers in the UK to question the necessity for a separate vent system, effectively re-ploughing Putnam's furrow. Wise (1952) considered whether careful application of the laws of fluid mechanics could offset the likelihood of trap seal depletion. Reducing the rate of outflow to an appliance branch limited self-siphonage, while limits on overall discharge linked to stack diameter would assist in reducing induced siphonage

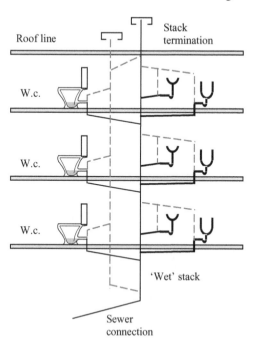

Figure 1.6 One-pipe system introduced in the 1930s

and back pressure trap seal losses. This fundamental research led directly to the single stack system, Wise and Croft (1954), Figure 1.7, initially applied to low-rise buildings but later extended up to 30 storeys.

A hybrid modified one-pipe system was introduced to extend the advantages of the single stack system, Figure 1.8, where the wet stack is connected to a parallel vent stack; there are no direct vent connections to appliances thus reducing the complexity of the networks and facilitating internal positioning of the drainage network in high-rise buildings. The single stack and modified one-pipe systems form the basis for both UK and Australian codes while the one-pipe system remains in use in the USA and mainland Europe. An equivalent single stack system introduced into the US ASPE (2002) design guidance is known as the 'Philadelphia' system, however this still requires local approval. While the traditional titles for these drainage options have been used in Figures 1.5 to 1.8, the more recent terminology is included in Chapter 8.

The range of trap seal depletion events to be prevented by the operation of the vent system is illustrated in Figure 1.9. Evaporation, Figure 1.9(a), is caused by the local ambient conditions. The routine installation of floor drains in plant rooms is now not advisable as the modes of floor cleansing no longer provide the water needed to 'top up' the trap. It will be shown later that the SARS spread in Hong Kong's Amoy Gardens housing complex was aided by the presence of dry floor traps. The vent system cannot prevent these losses.

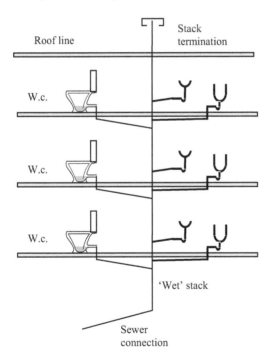

Figure 1.7 Single stack drainage system introduced in the UK in the 1970s

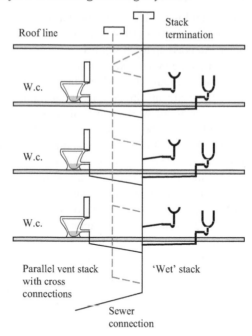

Figure 1.8 Modified one-pipe system featuring cross-ventilation

Self-siphonage, Figure 1.9(b), is caused by the appliance discharge having sufficient momentum to carry the trap seal out into the connected branch. Wise and Croft (1954) recommended either increasing the branch diameter downstream of the trap, this is now standard practice, or providing a local vent relief by a connection to the vent stack. In more recent times this recommendation may be met by providing a local air admittance valve.

Induced siphonage, Figure 1.9(c), is due to the air pressure transients propagating within the drainage network and may be avoided by local venting or careful selection of the branch diameter and slope in single stack applications. The maximum water flows to be carried by the drainage network also determines the likely range of pressures applied to the trap seal and are limited by codes – stack diameters increase as the expected system loading rises.

Back pressure, Figure 1.9(d), is due to positive air pressure transients generated within the system either by system surcharge, at the stack base or at any stack offset, or by positive pressures entering the network from the sewer, again possibly due to a remote system surcharge or pump operation. The solution is to vent the branch leading to the trap or to provide a flexible containment device locally to absorb the incoming positive transient – this option will be discussed and demonstrated in later chapters.

Wind-driven oscillation of the trap seal may occur due to the wind shear over the roof level stack termination and may lead to visible trap seal oscillation and possible depletion and may be avoided by 'shading' or capping the vent

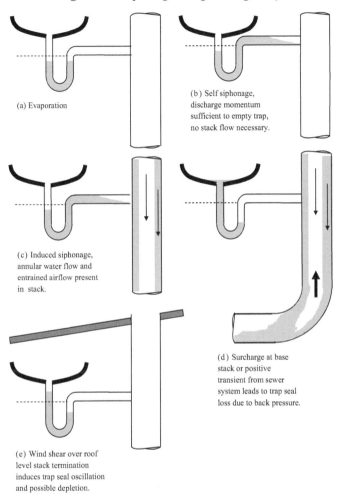

(a) Evaporation

(b) Self siphonage, discharge momentum sufficient to empty trap, no stack flow necessary.

(c) Induced siphonage, annular water flow and entrained airflow present in stack.

(d) Surcharge at base stack or positive transient from sewer system leads to trap seal loss due to back pressure.

(e) Wind shear over roof level stack termination induces trap seal oscillation and possible depletion.

Figure 1.9 Trap seal loss due to the pressure regime within the building drainage and vent system

termination. Therefore the majority of trap seal depletion may be prevented by vent system design to control the pressure regime within the network. While historically the design advice was based on extrapolation of limited laboratory investigations, and some site testing, leading to a codified approach not based on any fundamental analysis of the transient nature of the phenomena considered, modern simulation techniques allow a reassessment of venting design, as will be demonstrated in later chapters.

Figure 1.10 presents a particular view of the development of building drainage and vent systems. As mentioned above, the early Victorian obsession with odour ingress prevention, as well as the miasma theory of infection spread supported by

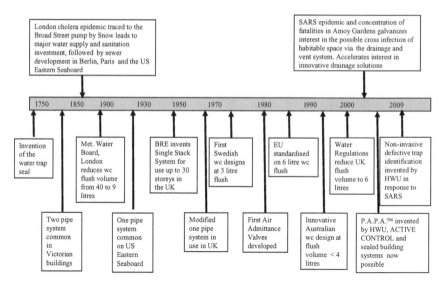

Figure 1.10 Time line for the development of drainage and vent systems between two major public health events, the cholera epidemic of 1855 and the SARS event in 2003

Chadwick, Finer (1952), led to the complex venting and separate black and grey water system designs. However the cholera epidemic of 1855, leading to Snow's work on establishing the waterborne nature of this disease, by identifying the Broad Street pump, Hempel (2007), as the source of the local infection, led to a major reassessment and eventually to the major London sewer refurbishment by Bazelgette, Cook (2001). Chadwick's concerns took a secondary role. However the 2003 SARS epidemic, and in particular the spread of the disease within the Amoy Gardens housing complex in Hong Kong, once more reinforced concerns as to the airborne spread of disease as it was shown that cross-infection via airborne virus transmission, through dry traps, was a major contributory factor to the eventual death toll. In some respects then Chadwick was right. Figure 1.10 presents the development of the modern drainage and vent system as a process 'bookended' by the Broad Street cholera outbreak and the SARS epidemic in Hong Kong. Both events galvanised research and development into the process involved in drainage and vent system operation and emphasised the necessity for a clear understanding of system operation and potential failures.

The necessity to retain appliance trap seals during the operation of the building drainage system led to a body of research, from the 1930s onwards, directed at providing design guidance that would link acceptable water flow capacity to the resultant air pressure excursions within the network and hence to trap seal retention. The earliest work is due to Hunter (1924) at the National Bureau of Standards (NBS), Washington DC, who established the now accepted limit for stack water annular flow – namely an annular film whose cross-section is no more than 25 per cent of the stack cross-section. In terms of film thickness this implies a limit of 1/16 of the stack diameter. Similarly the acceptance of 50 or 75 mm water seal traps limited the pressure excursions to be tolerated within the network to ±375 N/m² or 37 mm water gauge. It will be shown later that this is the equivalent of a rapid change in airflow velocity of around 1 m/s.

During the 1950s Wise and his co-researchers at the UK Building Research Establishment developed the single stack system that reduced the venting complexity. Research continued into the 1970s, leading to the modified one-pipe system and included site as well as laboratory testing, Pink (1973b). Pink and Wise's work on stack length effects was important as it will be seen later that it drew directly on the application of the steady flow energy equation to the stack entrained airflow and challenged the concept that air entrainment was solely based on the appliance discharge flow with no acknowledgement of the system resistances.

In the United States Hunter's work was continued by Wyly and Eaton (1961), Wyly (1964), Wyly and Sherlin (1972), and Wyly and Galowin (1985) who identified relationships between the applied laboratory steady flow and the resulting stack pressure regime. This work was however predicated on the need to retain a complex venting network. Simplification of the venting system was also the prime objective of continuing work at the National Bureau of Standards that featured both laboratory and site trials on reduced venting installations, Galowin and Kopetka (1982), Galowin and Cook (1985). Research from the National Bureau of Standards also informed more recent discussions as to the form of European codes, particularly in the area of allowable maximum water flow and its relationship to stack diameters where a range of conflicting recommendations, from 25 to 33 per cent of stack cross-section, featured in the individual national codes, de Cuyper (1993).

During the 1980s a new concept in vent system design was initiated by the introduction of the first air admittance valves in Scandinavia. Positioning AAVs adjacent to an appliance trap seal effectively vented the appliance branch as soon as the air pressure fell below a preset level. This form of locally venting removed the necessity for long vent connections to the roof termination of the vent or wet stack. The air admittance valve concept is drawn from the wider field of surge protection involving inwards relief valve operation and can be modelled satisfactorily, Swaffield and Campbell (1992b). AAVs have slowly gained acceptance, the major issue in code terms being concerns as to the self-sealing capability of the diaphragm and therefore the potential for cross-contamination.

Up to the 1990s there was still no analytical, as opposed to experimental extrapolation based on steady flow conditions, methodology to aid vent system design. Wise (1957) had investigated the local pressure losses imposed on the entrained airflow during its passage through the drainage and vent system and Lillywhite and Wise (1969) for the first time introduced the steady flow energy equation applied overall across the system, from airflow entry through the roof vent termination to eventual passage into the sewer system. Although this was only demonstrated for steady flow it was an important first step in the application of established fluid mechanics analysis to the drainage network. A reappraisal of the mechanisms of air entrainment and a linking of the pressure recovery in the wet stack, due to the water-to-air interface shear forces, led to the introduction of a pseudo-friction factor approach that for the first time linked annular water downflow to the entrained airflow, Jack (1997), Swaffield and Jack (1998) and Jack (2000). Jack essentially provided an approach that allowed the pressure recovery experienced in the 'wet' stack, due to the shear forces acting between the annular water film and the entrained air core, to be represented in a manner analogous to the standard frictional resistance model, thus facilitating the introduction of a 'pseudo-friction factor' as a driver in the equations representing overall system operation. This work opened up the possibility of relating the air pressure regime within the network to changes in appliance discharge water flow and hence provided a direct simulation route that predicts trap seal retention for a wide range of system operational profiles. These developments will be discussed in detail later in this book.

In parallel to the simulation development mentioned, the more traditional drainage and vent system research continued, with major efforts in Taiwan where Cheng et al. (2008) reported fundamental work on the air pressure regime in full-scale drainage networks in both Japan and Taiwan. Similar work in Hong Kong addressed the issues of overcrowding on the actual rather than the assumed water flows discharged to the drainage network in high-rise buildings, Wong and Mui (2009), while Fernandes and Gonçalves (2006) investigated the application of AAVs within the Brazilian national codes.

The introduction of the AAV in the 1980s provided the first step in the next reduction in venting complexity as the local introduction of an AAV was capable of suppressing local negative pressure surges and hence protecting the appliance trap seal from induced siphonage. The possibility of back pressure trap seal depletion was not addressed, apart from a reliance on the vent system connections, until the variable volume containment device designed to deliver positive transient attenuation (PTA), invented by Swaffield and Campbell at HWU during 2000, Swaffield (2003), and developed by Gormley (2002), was introduced in 2003 by Studor Ltd as the P.A.P.A.™, following a research programme that recognised the linkages to pressure surge theory and air chamber design, Swaffield, Campbell and Gormley (2005a, 2005b). Taken together with the AAV, the introduction of a variable volume containment device allows the introduction of an Active Control strategy where both positive and negative air pressure transients may be suppressed and leads directly to the proposal for a 'sealed building drainage and vent system', Swaffield (2006), a proposal initially developed to aid in the design of drainage networks for security sensitive buildings but having its first application in the O_2 'Dome' venue in London's docklands.

Vent system research continues to be an active area and includes, as a direct consequence of the SARS epidemic, efforts to develop non-invasive and non-destructive methodologies to identify appliance trap seal depletion as a positive response to the cross-infection identified during that epidemic, Swaffield (2005), Kelly et al. (2008a, 2008b, 2008c).

Basic system design is still based on national design codes, however there are now notable examples where the application of various individual codes to the same building design results in quite different results and recommendations. Clearly this can only be explained in terms of shortcomings in the code interpretation of the actual mechanisms of drainage system operation, White (2008), as clearly the determining laws of physics do not recognise national code boundaries. There are therefore still several areas of research that require attention before drainage system design can be fully accepted as having a theoretical and analytical base.

1.4 Air entrainment, annular flow and terminal velocity

Experimental observations as early as Hunter (1924) established that the discharge water flow within the vertical wet stack took the form of an annular flow once the initial flow disruption due to the branch to stack entry abated. As illustrated in Figure 1.2(c) the flow below the junction will have a high degree of swirl as it establishes the annular film. More recently, Campbell and MacLeod (1999) also suggest that droplets of water continue to fall within the entrained air core, a result at variance with the earlier results, Pink (1973a), which failed to detect any water droplets in the air core. Several properties of the annular flow are worthy of further clarification, in particular the establishment of a terminal annular water velocity, a definition of the distance fallen before the terminal velocity is reached and an assessment of the air entrainment.

Figure 1.11 illustrates the development of the annular water film within the wet stack below a discharging junction. Early researchers were surprised to discover that the annular water flow in the vertical wet stack quickly attained a terminal velocity defined by the water flow rate and the dimensions and properties of the stack. Consider the annular flow illustrated by Figure 1.11.

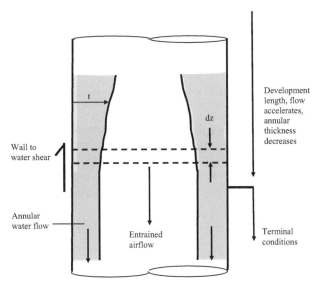

Figure 1.11 Annular flow development in a vertical stack leading to the establishment of a terminal velocity

Application of Newton's second law yields for a vertical element of the flow dy

$$\rho g(\pi D t dz) - \tau_0 \pi D dz = \rho \pi D t dz \frac{dV_w}{dt}, \qquad (1.1)$$

where t is time. If the interface shear between the water and air is neglected, Equation 1.1 reduces to

$$\rho g t - \tau_0 = \rho t \frac{dV_w}{dt}, \qquad (1.2)$$

Under terminal flow conditions the water acceleration is reduced to zero so that Equation 1.2 becomes

$$\tau_0 = \rho g t \qquad (1.3)$$

The wall to annular water flow shear stress may be expressed in terms of an empirical friction factor as

$$\tau_0 = \frac{1}{2} \rho f V_w^2 \qquad (1.4)$$

where f is the Darcy friction factor, so that the terminal thickness, t, and water terminal velocity, V_t, are linked as

$$V_t = \sqrt{\frac{2gt}{f}} \qquad (1.5)$$

Note that V_w becomes V_t once terminal conditions have been reached.

The friction factor, f, is most accurately represented by the Colebrook–White equation modified to accommodate free surface flow conditions, Bridge and Swaffield (1983), so that from Equation 1.5 it follows that

$$\frac{1}{f} = \frac{V_w}{\sqrt{2gt}} = \frac{Q_w}{(\pi D t \sqrt{2gt})} = -4\log\left(\frac{k}{14.84m} + \frac{0.315}{\mathrm{Re}\sqrt{f}}\right) \qquad (1.6)$$

where m is the free surface flow hydraulic mean depth, defined as (area A / wetted perimeter P), k is the wall surface roughness and Re is Reynolds number, defined as $\mathrm{Re} = \rho V_w m / \mu$. The hydraulic mean depth for an annular flow is thus $m = \pi D t / \pi D = t$ and so the Reynolds number may be expressed as $\mathrm{Re} = \rho V_w t / \mu = \rho Q_w / (\mu \pi D)$ and the Colebrook–White expression, Equation 1.6, becomes

$$\frac{Q_w}{(\pi D t \sqrt{2gt})} = -4\log\left(\frac{k}{14.84t} + \frac{0.315\mu}{\rho t \sqrt{2gt}}\right) \qquad (1.7)$$

Equation 1.7 may be used to determine terminal thickness, t, and hence terminal velocity, $V_t = Q_w / (\pi D t)$, and may be used to define the limits to stack flow if the assumption that only 25 per cent of the stack cross-sectional area can be taken up by the water annulus, so that t_{max} = D/16. Values of acceptable stack flowrates will be discussed in Chapter 8 in the context of national design codes.

In order to express terminal velocity in terms of water flowrate, rather than velocity, substitute

$$V_t = \frac{Q_w}{\pi D t} \tag{1.8}$$

into Equation 1.3 to yield

$$\tau_0 = \frac{1}{2}\rho f V_t^2 = \rho g t = \rho g \frac{Q_w}{\pi D V_t} \tag{1.9}$$

reducing to

$$V_t = \sqrt[3]{\frac{2g}{f} \frac{Q_w}{\pi D}} \tag{1.10}$$

The term $2g/f$ may be recognised from the Chezy equation for steady flow in an open channel as a definition of the Chezy resistance coefficient, C, that is often defined in terms of the flow hydraulic mean depth m, where m = (flow cross-sectional area / wetted perimeter), Douglas et al. (2005), which reduces as shown to t for annular flow, and the Manning open channel resistance coefficient, n, by the expression

$$C = \sqrt{\frac{2g}{f}} = \frac{m^{1/6}}{n} = \frac{t^{1/6}}{n} = \frac{1}{n}\left(\frac{Q_w}{\pi D V_t}\right)^{\frac{1}{6}} \tag{1.11}$$

Substitution in Equation 1.10 yields

$$V_t = \sqrt[3]{\frac{1}{n^2}\left(\frac{Q_w}{\pi D V_t}\right)^{\frac{1}{3}}\frac{Q_w}{\pi D}} \quad \text{so that} \quad V_t^{\frac{10}{9}} = \sqrt[3]{\frac{1}{n^2}\frac{1}{\pi^{4/3}}\left(\frac{Q_w}{D}\right)^{\frac{4}{9}}} \tag{1.12}$$

reducing to an expression for terminal velocity in terms of stack roughness and applied water flowrate

$$V_t = K\left(\frac{Q_w}{D}\right)^{0.4} \tag{1.13}$$

provided that $K = \left(\frac{0.2173}{n^2}\right)^{0.3} \tag{1.14}$

This expression is helpful and was first developed by Wyly (1964). A typical smooth pipe value of Manning n of 0.007 yields a K value of 12.4 which compares well with Wyly's initial value, for cast iron pipes from 75 to 150 mm in diameter, of 7.72 in SI units, and later work by Chakrabarti (1986). Results from the more comprehensive Colebrook–White expression (1.7) suggest a value of K=14.99, Swaffield and Galowin (1992), for smooth pipes with k=0.

Thus terminal velocity is a feature of stack annular flow which suggests that for a steady or long duration downflow the shear stress induced entrained airflow will be subject to a constant driving force.

The vertical distance required to achieve terminal velocity is also an important consideration if this model of air entrainment is to be developed further. In tall buildings it would be helpful if the distance required was relatively short, thus facilitating a simpler model of the shear driving force.

$$\frac{dV_w}{dt} = \frac{dV_w}{dz}\frac{dz}{dt} = V_w\frac{dV_w}{dz}$$

so that, as from Equation 1.2, $\dfrac{dV_w}{dt} = g - \dfrac{\tau_0}{\rho t}$, substitution yields

$$\frac{dV_w}{dz} = \frac{1}{V_w}\frac{dV_w}{dt} = \frac{1}{V_w}\left(g - \frac{1}{2}f\frac{\pi D}{Q_w}V_w^3\right)$$

and substituting for V_t from Equation 1.10 yields

$$dz = \frac{V_w dV_w}{\left(g - \dfrac{1}{2}f\dfrac{\pi D}{Q_w}V_w^3\right)} = \frac{V_t^2(V_w/V_t)d(V_w/V_t)}{\left(g(1 - \dfrac{1}{2g}f\dfrac{\pi D}{Q_w}V_w^3)\right)} = \frac{V_t^2(V_w/V_t)d(V_w/V_t)}{\left(g(1-V_w^3/V_t^3)\right)}$$

(1.14)

So that if $\theta = V_w/V_t$

$$\int_{Z=0}^{Z=Z_T} dz = \frac{V_t^2}{g}\left(\frac{\theta d\theta}{1-\theta^3}\right)$$

(1.15)

The result of this integral will be infinite as the velocity will approach its terminal value asymptotically, however normal practice is to set $\theta = 0.99$ as the upper integration limit so that the vertical distance to terminal velocity becomes

$$Z_t = 0.159V_t^2$$

(1.16)

Terminal lengths rarely exceed two storeys and, as will be shown in Chapter 8, are often less than 5 m.

The measurement of terminal velocity remains an ongoing line of research. Cheng et al. (2009) challenged the Wyly predictions of terminal velocity as being too low and presented data based on video capture of the stack annular flow used to determine the film velocity. However it is difficult to identify the flow element being measured as there is evidence of roll waves within the vertical stack annular flow, as first identified by Jack, Swaffield and Filsell (2004), Filsell (2006). Figure 1.12 presents laser Doppler anemometry results for a limited series of annular downflow in a glass 100 mm diameter vertical stack. It will be seen that the results lie between the $n=0.007$ and $k=0$ assumptions of smooth pipe annular flow resistance and, from the second x-axis, $(Q_{water})^{0.4}$, that the general form of Wyly's expression is confirmed for a single diameter stack, Thancanamootoo (1991),

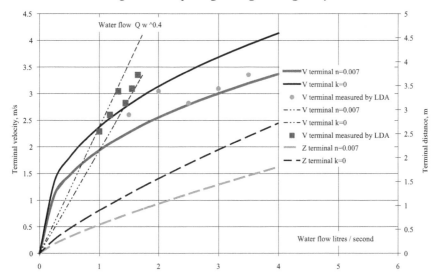

Figure 1.12 Terminal velocity measurements in a 100 mm diameter glass vertical stack cf. predictions based on Equation 1.12; terminal distance predictions are also included

Swaffield and Thancanamootoo (1991). Thancanamootoo also found that at the higher flowrates the prediction of annular thickness compared favourably with the LDA results obtained by traversing the focus of the laser beams until the signal broke due to change of fluid from water to air.

The annular flow assumed within vertical stacks is far from the uniform film that might be imagined based on published work. There is evidence of non-uniformity in the vertical direction even if the applied water downflow is steady, the observed annular flow demonstrates characteristic roll wave as found in free surface flows on dam spill ways or during heavy rain on steep gradient roadways, Chow (1959), Brock (1970), Patnaik and Perez-Blanco (1996). Figure 1.13 illustrates the effect and implies that measurements of annular mean flow velocity is at best problematic. In practice the duration of any particular downflow and the rate of change of the water flow is likely to minimise the effect of such non-uniformities.

These expressions for terminal velocity and terminal distance encouraged earlier researchers to attempt to link these values to entrained airflow, Wyly and Eaton (1961), and it is useful to follow the development of a proposed relationship dependent upon the annular flow area to stack cross-section ratio, r_s and stack design, based on the following expressions derived or stated earlier.

$$r_s = \frac{\pi D t}{\pi D^2 / 4}, \quad Q_w = \pi D t V_t, \quad V_t = K\left(\frac{Q_w}{D}\right)^{0.4}$$

Hence

$$r_s = \frac{4 Q_w}{\pi D^2 V_t} = K_1 \frac{Q_w^{0.6}}{D^{1.6}} = K_1 \left(\frac{Q_w}{D^{8/3}}\right)^{3/5} \tag{1.17}$$

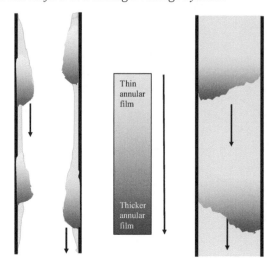

Figure 1.13 Roll waves on the annular stack water downflow are seen as localised increases in depth angled across the stack as shown

where K_1 is a constant. The entrained airflow Q_a is given by $Q_a = V_a A_a$ where A_a is the air core cross-section, $(1 - r_s)\pi D^2 / 4$ and V_a is a mean entrained airflow velocity at that point in the vertical stack. Wyly proposed that the air velocity was related to the terminal water velocity $V_a = CV_t$ where the value of the constant C was assumed to be unity. It therefore follows that an expression for entrained airflow may be developed as

$$Q_a = CV_t \left(1 - r_s\right)\pi D^2 / 4 = CK \left(\frac{Q_w}{D}\right)^{2/5} \left(1 - r_s\right)\pi D^2 / 4$$

Collecting all constants and substituting for Q_w from Equation 1.17 allows a relationship to be proposed that links entrained airflow to stack properties and an assumption as to the maximum allowable water flow in terms of the annular area to stack cross-section ratio,

$$Q_a = K_2 r_s^{2/3} (1 - r_s) D^{8/3} \tag{1.18}$$

The corresponding water flow determined by setting the value of $r_s = 25$ per cent is given, Wyly and Eaton (1961), from Equation 1.18 as

$$Qw = K_1 r_s^{5/3} D^{8/3} \tag{1.19}$$

Wyly and Eaton (1961) found the value of K_1 to be 31.9 for a cast iron pipe and with Q in m³/s and D in m. The water flow may be derived as $1.6D^{8/3}$ for a stack flowing 1/6 full and $3.15D^{8/3}$ for a stack flowing 1/4 full.

Wyly and Eaton (1961) presented these expressions which are based on a fixed ratio between the annular water flow terminal velocity and the entrained airflow mean velocity. It takes no account of the effect of pressure gradients identified by Wise (1986) nor of the earlier experimental work by Pink (1973b). Basically this expression is fundamentally flawed as it ignores the basic steady flow energy equation that links the shear forces available in the wet stack to the pressure losses imposed on the entrained airflow as it passes through the various restrictions to flow already identified in the vertical stack, Wise (1986), Jack (2000), and will be returned to in Chapter 8 in a discussion of the air entrainment values included in national codes.

1.5 Assessment of system water flows – fixture and discharge unit approaches

It will be appreciated that the driving function for all the low amplitude air pressure transients propagated within a building drainage and vent system is the annular water flow in the system vertical stacks, its time dependency and its interaction with the geometry of the network. This is equally true for the negative transients propagated as a result of an increased water downflow following appliance discharge as it is for the positive air pressure transients generated by system surcharge and interruptions to the airflow path.

Therefore the assessment of system usage has been a major research effort over the past 50 years. Early work at BRE in the UK, Webster (1972), identified the main sanitary appliance usage periods during the day for domestic situations – namely 06.00–09.00, 17.00–20.00 and 21.00–24.00, with sinks being more uniformly distributed. More recent work by Butler (1991, 1993), Edwards and Martin (1995) and Butler and Davies (2004) generally confirmed these results. Historically water usage studies have also been undertaken by Konen (1989), Kiya (1977) and Murakawa (1989).

The random appliance usage leads inevitably to the need for a probabilistic approach to the likely water flows to be found in building drainage networks at any time. Wise and Croft (1954) considered the maximum hourly probabilities for the discharge of w.c.s, sinks and washbasins in domestic usage based on a model where the probability, p, of the fixture being in use was determined as $p=t/\text{T}$, where t is the duration of appliance discharge and T is the interval to the next usage. It also follows that the probability of this appliance not being in use is $(1 - p)$. The results of this work, summarised in Table 1.1 have been extensively used and are worth noting here.

Table 1.1 Maximum hourly probabilities for appliance discharge in domestic use, Wise (1986)

	Duration of discharge, t seconds	Interval between discharges, T seconds	$p = t/T$
W.c.	5	1140	0.0044
Washbasin	10	1500	0.0067
Sink	25	1500	0.0167

The intervals between appliance usage obviously vary dependent upon the type of installation considered. While the domestic frequency may be based on a 1200 second interval it is clear that in congested applications, such as airport washrooms, the toilet use frequency may well be as short as the refill time of the cistern – perhaps 2 minutes. Generally UK codes have adopted the work of Wise and Croft and the time intervals are often presented as domestic at 1200 seconds, 600 seconds for public or commercial and 300 seconds for congested usage.

Thus the water flow rate at any one time depends on the appliances discharging within a particular time period related to the transit time of a discharge through the network. For example a bath discharging to a network will set up a long time water flow which is then reinforced by any shorter duration discharges. It is clear that a determination of the likely appliance discharges present in any time increment has to be approached from a statistical perspective unless simulation packages are available that allow user defined appliance discharge to be considered in real time. The first and most well-known contribution to the development of a statistical approach to water flow estimation was due to Hunter (1940) working at the National Bureau of Standards in Washington DC. His work remains the basis of the main US codes and has influenced other national codes. Effectively the resulting flow estimation yields a quasi-steady flow value that may be used to size the drainage and vent network based on other experimental results. However Hunter understood clearly the limitations of his approach when, in his 1940 paper, he stated that

> ... the conventional pipe formulae apply to the irregular and intermittent flows that occur in plumbing systems only during that time (usually very short) and in that section of pipe in which the variable factors involved (velocity or volume rate of flow or hydraulic gradient and hydraulic radius) are constant ...

implying that he recognised that the treatment of the unsteady flow conditions was necessary but that the necessary mathematical simulation tools were not available.

Wise and Croft (1954) introduced a similar approach, later also taken up by Griffiths (1962) and Burberry and Griffiths (1962). Wise, in Wise and Swaffield (2002) presents the clearest explanation of the development of the probabilistic approach to flow estimation represented by these contributions.

The probability of any number of a particular appliance type discharging simultaneously is found as the product of individual probabilities, see Table 1.1, the probability of n being simultaneously in use is p^n. Application of the binomial theorem allows the probability, P, that any number, r, of appliances out of a total number, n, of the same sort will be given by

$$P = c_r^n p^r (1-p)^{n-r} \tag{1.20}$$

where the term c_r^n is given by $c_r^n = \dfrac{n!}{r!(n-r)!}$ and $n!$ denotes factorial n defined as $n! = n(n-1)(n-2)...$

Equation 1.20 may be reduced to the Poisson expression if the individual probabilities are small so that

$$P = \frac{e^{-\varepsilon}}{r!} \varepsilon^{r}$$ (1.21)

Consider a network containing 10 each of w.c.s, sinks and washbasins. If a criterion is set that actual loadings on the system should exceed the design load for less than 1 per cent of the time then it is possible to deduce the number of any one type of appliance that may be allowed to discharge simultaneously within this limitation. The design load may be identified as the number of appliances in simultaneous use with a probability close to 0.01. Table 1.2 demonstraits this criterion applied to each of the three appliance types and each of the levels of usage, domestic, commercial and congested. Based on this criterion the design load under congested conditions would be 2 w.c.s, 2 washbasins and 3 sinks, these figures reducing to 1 w.c., 1 washbasin and 2 sinks in a domestic situation. This is intuitively correct as it represents a lower design flowrate in the domestic application.

In general terms therefore the design load is represented by an expression of the form

$$\sum_{r=r}^{r=n} c_{r}^{n} p^{r} (1-p)^{n-r} = 0.01$$ (1.22)

The discussion above does not recognise that appliance discharges may overlap and may therefore be expected to overestimate the applied flow. Hunter recognised this and was the first to introduce the concept of the discharge unit, effectively a codification that represents the appliance's load producing contribution. Wise and Swaffield (2002) again present a summary of the development of this procedure that is now found in some form in all national codes. The discharge unit method relies on the observation that the same overall flowrate may be derived from the usage of differing numbers of different appliances. Wise gives the historic UK example that an overall flow of 10 l/s may be generated by 160 w.c.s, each with a 13 litre flush volume – a standard flush volume in Scotland prior to the 1980s – or 270 sinks. Hence the relative weighting of the w.c. to the sink is 270:160. If the w.c. is allocated the purely arbitrary discharge unit value of 10 then the sink has a relative fixture unit value of 5.9. In order for this methodology to be robust, Wise identifies the following prerequisites:

1 a carefully chosen set of units to express the load producing contribution of each different appliance used at differing frequencies,
2 a relationship between flowrate and total appliance loading expressed in terms of the fixture unit allocations.

These two criteria are linked. The arbitrary nature of the base considerations and allocations has historically led to the discharge unit method being poorly understood and has contributed to a widening gap between designers wedded

Table 1.2 Design load determination based on the probability of usage of individual appliances. Note n is the total number of any one appliance type and r is the number in simultaneous use, t is the duration of flow from an appliance and T is the interval between uses

Appliance	n	t seconds	T seconds	p=t/T	1-p	r=0	r=1	r=2	r=3	r=4	r=5	r=6	r=7	r=8	r=9	r=10
W.C.	10	5	300	0.01667	0.98333	0.845294	0.14327	0.0109274	0.000494	1.465E-05	2.9795E-07	2.6556E-08	1.2646E-10	5.2691E-13	9.7575E-16	1.6538E-18
W.C.	10	5	600	0.00833	0.99167	0.919723	0.077288	0.0029226	6.55E-05	9.631E-07	9.7123E-09	4.1846E-10	9.9632E-13	2.0757E-15	1.9219E-18	1.6151E-21
W.C.	10	5	1200	0.00417	0.99583	0.959106	0.04013	0.0007556	8.43E-06	6.173E-08	3.0994E-10	6.5658E-12	7.8165E-15	8.1442E-18	3.7695E-21	1.5772E-24
Washbasin	10	10	300	0.03333	0.96667	0.712471	0.24568	0.0381227	0.003506	0.0002115	8.7534E-06	1.6708E-06	1.5912E-08	1.326E-10	4.9112E-13	1.6935E-15
Washbasin	10	10	600	0.01667	0.98333	0.845294	0.14327	0.0109274	0.000494	1.465E-05	2.9795E-07	2.6556E-08	1.2646E-10	5.2691E-13	9.7575E-16	1.6538E-18
Washbasin	10	10	1200	0.00833	0.99167	0.919723	0.077288	0.0029226	6.55E-05	9.631E-07	9.7123E-09	4.1846E-10	9.9632E-13	2.0757E-15	1.9219E-18	1.6151E-21
Sink	10	25	300	0.08333	0.91667	0.418904	0.380822	0.1557907	0.037767	0.0060085	0.00065547	0.00038681	9.2097E-06	1.9187E-07	1.7766E-09	1.6151E-11
Sink	10	25	600	0.04167	0.95833	0.653380	0.284078	0.0555805	0.006444	0.0004903	2.5582E-05	6.3186E-06	7.5221E-08	7.8355E-10	3.6276E-12	1.5772E-14
Sink	10	25	1200	0.02083	0.97917	0.810151	0.172373	0.0165038	0.000936	3.487E-05	8.9018E-07	1.0087E-07	6.0044E-10	3.1273E-12	7.2391E-15	1.5402E-17

Appliance	n	t seconds	T seconds	p=t/T	1-p	0 or more	1 or more	2 or more	3 or more	4 or more	5 or more	6 or more	7 or more	8 or more	9 or more	10 or more
W.C.	10	5	300	0.01667	0.98333	1.00	0.154706	**0.0114362**	0.000509	1.497E-05	3.2464E-07	2.6683E-08	1.2699E-10	5.2788E-13	9.7741E-16	1.6538E-18
W.C.	10	5	600	0.00833	0.99167	1.00	**0.080277**	0.0029891	6.65E-05	9.733E-07	1.0132E-08	4.1945E-10	9.984E-13	2.0776E-15	1.9235E-18	1.6151E-21
W.C.	10	5	1200	0.00417	0.99583	1.00	**0.040894**	0.0007641	8.49E-06	6.205E-08	3.1651E-10	6.5737E-12	7.8246E-15	8.1459E-18	3.7711E-21	1.5772E-24
Washbasin	10	10	300	0.03333	0.96667	1.00	0.28753	**0.0418502**	0.003728	0.000222	1.044E-05	1.6868E-06	1.6045E-08	1.3309E-10	4.9281E-13	1.6935E-15
Washbasin	10	10	600	0.01667	0.98333	1.00	0.154706	**0.0114362**	0.000509	1.497E-05	3.2464E-07	2.6683E-08	1.2699E-10	5.2788E-13	9.7741E-16	1.6538E-18
Washbasin	10	10	1200	0.00833	0.99167	1.00	**0.080277**	0.0029891	6.65E-05	9.733E-07	1.0132E-08	4.1945E-10	9.984E-13	2.0776E-15	1.9235E-18	1.6151E-21
Sink	10	25	300	0.08333	0.91667	1.00	0.58144	0.2006183	**0.044828**	0.0070601	0.00105168	0.00039621	9.4034E-06	1.9366E-07	1.7927E-09	1.6151E-11
Sink	10	25	600	0.04167	0.95833	1.00	0.346625	**0.062547**	0.006966	0.0005223	3.1976E-05	6.3946E-06	7.6008E-08	7.872E-10	3.6433E-12	1.5772E-14
Sink	10	25	1200	0.02083	0.97917	1.00	0.189849	**0.017476**	0.000972	3.586E-05	9.9166E-07	1.0148E-07	6.0358E-10	3.1346E-12	7.2545E-15	1.5402E-17

Note – Congested, commercial and domestic usage intervals taken as T = 300, 600 and 1200 seconds respectively. Number of appliances identified correspond to an actual loading that does not exceed the design load for more than 1% of the time, integer number of appliances imply some degree of latitude.

to the 'codes' and researchers attempting to rationalise the flow loading and the resultant codes. These issues will be returned to in Chapter 8.

In order to determine the likely design load to be catered for by the drainage and vent system, the procedure is to sum the fixture unit values for the appliances connected to the network. The flowrate may then be deduced from graphical or tabular data together with coefficients designed to replicate the intensity of usage within a building type of any particular appliance. In addition, codes assign pipe sizes and slopes based on the likely number of discharge units. Hence the slope or diameter of a branch may increase if the combined discharge unit total being carried by that section of the network exceeds a preset value. As shown earlier, attempts have been made to determine the entrained airflow in any part of the network based on the discharge unit summation at that point in the system – this approach is severely limited as it takes no account of the system resistances that determine the entrained airflow in response to the wet stack shear forces. These design methods will also be discussed in Chapter 8.

1.6 Definition of airflow as unsteady, relationship to pressure surge theory

It is clear from the discussion above that the air entrainment within a building drainage and vent system depends upon the number and type of appliances discharging to the network as well as on both the interactions of that discharge water flow with the physical form of the drainage pipework and the impact upon the system air pressure regime of events remote from the system, including sewer surcharges, pump operation or wind induced or mechanical fan generated pressure fluctuations over any system open terminations. Appliance discharge introduces time dependency as the appliance operation will be random and the individual appliance discharge profile will be time dependent, more so in the cases of w.c.s than other appliances such as baths, basins or washing machines. Thus the water flows in the system are unsteady and hence the entrained airflow will be similarly time dependent.

The airflow within building drainage and vent systems thus becomes subject to the propagation of low amplitude air pressure transients, a normal consequence of any change in an airflow operating condition. Increases in water flow dependent upon appliance discharge will increase the entrained airflow and generate negative pressure transients while impediments to the entrained airflow, for example by surcharge of any portion of the flow network, will result in the propagation of positive pressure transients. Thus the time dependent air pressure regime within the drainage network as a result of system operation falls within that body of unsteady transient flow conditions generally referred to as pressure surge and solvable by well-established analysis and simulation techniques, albeit usually applied to flow conditions where a system failure may involve the propagation of excessive pressure surges, measured in atmospheres, whereas, as will be shown with reference to cross-contamination and the spread of infection, air pressure transients of 100 mm water gauge may be sufficient to lead to a fatally failed system.

In a divergence from the more usual pressure surge application, where transients are in general propagated as a result of changes in flow condition as a result of a change in a system boundary condition, for example a pump setting or valve operation, transients in the entrained airflow within building drainage and vent systems are propagated by both boundary conditions changes, such as system surcharge at a stack base or pressure fluctuations over a roof open vent, and changes in the annular water flow in the system's vertical stacks that entrain an airflow in the core as a result of the shear forces acting at the fluid interface. Thus the simulation of air pressure transients in building drainage and vent systems introduces the need for a wider range of driving functions than normally met in surge analysis.

However, having recognised that low amplitude air pressure transients in the entrained airflow are a normal consequence of system operation, it is necessary to develop the required equations to define the unsteady airflow and then to identify a simulation technique capable of sufficient flexibility to allow the analysis of varied systems with boundary conditions that will in many cases be dependent upon the local air pressure regime.

Later chapters will present both the detail development of the St Venant equations of continuity and motion, which are capable of defining the flow conditions discussed above, as well as the Method of Characteristics (MoC) numerical analysis technique that allows the transformation of the St Venant quasi-linear hyperbolic partial differential equations into a pair of total differential equations solvable by relatively basic finite difference techniques, Lister (1960). As with all pressure surge simulations the 'art' of the application of the MoC lies in the development of suitable boundary conditions, including in this application a representation of the shear interaction that initially entrains the airflow; suitable boundary conditions will be developed.

1.7   Time line for analysis techniques

Figure 1.14 illustrates the development of the application of the Method of Characteristics to low amplitude air pressure transient simulation within building drainage and vent systems and will form the basis of later discussion within this text. As will be shown in Chapter 2, the initial defining work on pressure surge or waterhammer analysis was due to Joukowsky (1900) working in St Petersburg at the end of the 19th century. The time dependency of the pressure surge problem meant that detailed simulation techniques capable of representing frictional losses as well as multiple and simultaneous boundary condition changes had to await the application of computing techniques in the 1960s, an effort led by Streeter and Lai (1962) in the USA and Fox (1968) in the UK, both of whom utilised Lister's timely exposition of the use of the Method of Characteristics to solve the St Venant equations. These techniques became the basis for the industry standard simulation of pressure surge and waterhammer and provided the basis for the extension of these techniques to the analysis of low amplitude air pressure transient propagation in building drainage and vent systems detailed in this book.

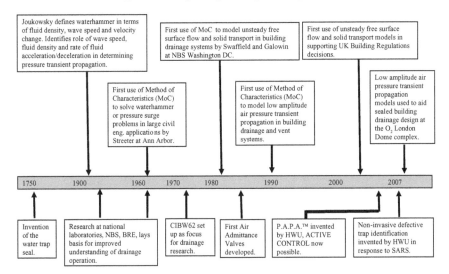

Figure 1.14 Development of the MoC applied to low amplitude air pressure transient simulation in building drainage networks

The application of MoC to drainage networks was initially directed towards the unsteady free surface flows found within building drainage networks as a result of random appliance discharge and wave attenuation, a research effort jointly based in the UK and at the National Bureau of Standards in Washington DC, Swaffield and Galowin (1992), that resulted in a simulation model DRAINET. The techniques developed have since been used extensively to model solid transport in attenuating flow, Swaffield and McDougall (1996), to assist in the development of UK Building Regulations, McDougall and Swaffield (2000), to illustrate the effects of water conservation on building drainage operation, McDougall and Swaffield (2003), and to define the effect of defective drainage on system operation, Swaffield, McDougall and Campbell (1999), as well as to investigate any low flow system operational advantages as a result of introducing non-circular building drainage pipework, a concept based on the use in sewer networks of ovoid drain sections, Cummings, McDougall and Swaffield (2007).

The application of MoC to model air pressure transients in drainage networks was initiated in 1989 at Heriot Watt University, Swaffield (1989), and led to the development of a network simulation model, AIRNET, that utilises the MoC solution to the St Venant equations, as well as an extensive library of boundary conditions derived through a series of research council and industry research programmes over a twenty-year period. That research effort, supported by the UK EPSRC, involved some 25 person years of funded researcher support from EPSRC and, more recently as the developed models were applied to generating new appliances and design techniques, industry. This research programme generated eight doctoral theses and a body of research papers that form the basis for the presentation in this book.

As shown by Figure 1.14, the development of both DRAINET and AIRNET was undertaken with encouragement from the Counseil International du Batiment Working Commission W62 whose members provided an annual forum for discussion of the research in both areas and led to many essential interactions, particularly with committed industrial colleagues. These interactions in turn led to the development by the Heriot Watt research team of a series of concepts of direct application to controlling and suppressing transient propagation within building drainage networks, limiting the extent of venting necessary and introducing a non-invasive and non-destructive test methodology to ensure trap seal retention within complex buildings, a necessity belatedly recognised following the SARS epidemic of 2003. Each of these applications is discussed later in Chapters 5 to 7.

The scale of the pressure transient propagation may be appreciated by reference to Joukowsky's original expression for the transient pressure, Δp, generated by an instantaneous change in airflow velocity, Δu, in a fluid flow where the fluid density, ρ, and the acoustic wave speed, c, are known, namely

$$\Delta p = -\rho c \Delta u \qquad (1.23)$$

Figure 1.15 illustrates the transient pressure generated following a 1 m/s change in airflow velocity, the air density being taken as 1.3 kg/m^3 and the wave speed as 320 m/s, namely 41.6 mm of water gauge. The figure also demonstrates the airflow rates necessary to support this pressure transient as the vent diameter increases; note that it is airflow velocity rather than flowrate that determines the pressure surge. An instantaneous transient will be shown to be possible under some flow conditions, including surcharge of the stack base, or at a stack offset, due to excessive water downflows, possibly due to overcrowding of accommodation or undersizing of the drainage network provision. Joukowsky's relationship also defines the sign of the transient, as local airflow is increased the transient generated has a negative sign as it is necessary to induce an increased flowrate through the system. Similarly a local reduction in air velocity generates a positive transient as the airflow through the system has to be reduced.

Thus it is accurate to define the propagation of low amplitude air pressure transients within building drainage and vent systems, generated as a result of both internal changes in operating condition and the impact of changing conditions external to or remote from the network, as an addition to the family of unsteady flow conditions addressed via the St Venant equations of continuity and motion and subject to simulation via application of the Method of Characteristics. This book will fully explore that assertion and show that the design of building drainage and vent systems, as well as the development of codes, based on the 'laws of physics', rather than the 'rule of thumb' and national preference, may be informed by transient analysis. Similarly it will be shown that the full understanding of system operation, as provided by systematic simulation, allows the development of innovative design approaches, as well as equipment aimed at the control and suppression of transient propagation and techniques to identify potential trap seal depletion and avoid the possibility of cross-contamination health risks.

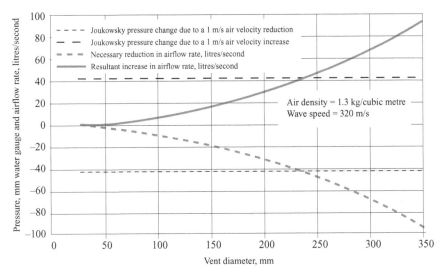

Figure 1.15 Relationship between the Joukowsky pressure transient generated by an instantaneous 1 m/s change in flow velocity, +ve or −ve, and the vent diameter

1.8 Concluding remarks

The current methodology for the design of building drainage and vent systems has developed and evolved over the past 150 years. Since the 1930s, it is clear that innovations in design have been research led, from the contributions of Hunter through the single stack innovations attributable to Wise to more recent developments such as the air admittance valve and positive air pressure attenuator. Despite this the national design codes are still firmly rooted in early experimental work based on steady flow conditions and surrogate values of appliance discharge, fixture units, that differ depending on which code is adopted. The introduction of an unsteady flow analysis, drawing heavily on industry standard approaches to pressure surge analysis, offers an alternative path when coupled to the capability of computing simulations to replace the more traditional estimates of system loading with user-defined profiles of appliance discharge sequences. There now exists the possibility that an analytical base can be developed that will allow the design of building drainage and vent systems to be approached with the rigour accepted across other building services system design and will allow solutions that are no longer dependent upon national preference rather than the laws of physics.

This book will explore the development of analytical solutions and the development of simulation techniques. In particular it will demonstrate that the perceived operation of drainage vent systems can be accurately modelled and that such modelling can be used to develop new and cost effective design solutions. The concepts of Active Control to suppress low amplitude air pressure transients and protect appliance trap seals will be demonstrated as will the underlying basis for a sealed building design methodology. The system simulations will also be used to highlight variations in national codes and to identify code recommendations that are not supportable, for example the relationship between wet stack and vent stack diameters or entrained airflows. Thus the analytical tools demonstrated offer an opportunity to develop unified codes – a long-held aspiration that has always foundered on national rules of thumb.

This book will demonstrate and propose the development of analytic tools based on proven applications of pressure transient theory, backed up by extensive laboratory and site testing that has the capability to bring the design of building drainage and vent systems up to the levels of accuracy commonly expected from other building services design.

2 Pressure transient propagation in building drainage and vent systems

While the purpose of a building drainage and vent system is to allow the rapid removal of liquid and solid waste from a building, the mechanisms inherent in system operation have been often regarded as in some way outside the application of quite standard fluid mechanics theory. While the flow conditions are accurately described as unsteady, i.e. responding to random appliance operation with time dependent discharge flow profiles, and multi-phase, i.e. the flow comprises free surface water flows, entrained airflows and intermittent solid transport, this does not imply that a theoretical base for system analysis is not available.

The complex building drainage and vent systems in large buildings will entrain airflows at rates many times the driving water flow rate and hence the mechanism by which air entrainment occurs, and the effect of changes in this water flow, determining air entrainment, become vital to an understanding of the pressure regime within the system and the impact that changes in pressure have on the survivability of the appliance traps seals that provide the protective barrier that minimises the risks of cross-contamination between the drainage network and habitable space.

This chapter will introduce the fundamentals of low amplitude air pressure transient propagation within building drainage and vent systems; the mechanisms of transient generation and propagation will be introduced together with a review of the necessary wave equations and their solution, initially in a frictionless system. Initial predictive techniques involving both the identification and quantification of transient reflection and transmission at system boundaries, both internal and terminal, will be introduced and the principle of superposition of pressure waves will be shown to allow prediction of the system pressure regime and to be an important tool in deciphering the transient pressure response of complex networks.

2.1 Development of pressure transient theory

Pressure transients are defined as the changes in local fluid pressure that propagate throughout a fluid network at the appropriate local acoustic velocity as a result of changes in the system operating point. Transients are generated whenever the current steady running conditions are modified, either as a result of an intended change in the system operating point, for example valve closure on completion of fuel transfer or an increased fan or pump speed to respond to some system control device, or an unexpected system failure, for example power failure to a pumping station leading to pump failure, column separation and possibly damaging pressure surge generation.

Thus pressure transient theory, historically referred to as waterhammer or pressure surge, is a condition relevant to all fluid system design and operation, ranging in its application from public utility water distribution networks, and sewage pumping mains, to long distance and sub-sea oil and gas transportation and specialist applications such as in-flight refuelling, fuel transfer and nuclear reactor cooling system operation. The more usual fluids considered are liquid and the pressure surges, measured in atmospheres or m of water gauge, propagating at wave speeds as high as 2000 m/s, have to be contained, controlled or designed out to prevent the significant risk of damage. Pressure transient propagation is also central to the communication of system operating point changes in gaseous systems, whether mechanical ventilation, fume cupboard extract networks or, as in the major applications to be dealt with in this specialist text, building drainage and vent systems. In these gaseous systems the pressure transients will be measured in mm of water gauge and will be propagated at the acoustic velocity in air, however the base theory will remain identical across all the applications mentioned and the analysis techniques all derive from the same historic international effort that has lasted over a century and has resulted in proven industry standard techniques that apply equally to building drainage and vent systems. This approach will offer a means of validating design codes against a central scientific base as opposed to rules of thumb, many nation based.

Acoustics research in the 19th century opened up the possibility that waterhammer – the damaging pressure surges generated by inadvertent system failure – could be approached from a scientific base. Early work by Korteweg (1878) established the value of wave propagation velocity, c, relative to the acoustic velocity in an unenclosed expanse of the flowing fluid, c_0, in an elastic tubular conduit that included the effect of conduit elasticity, E, diameter, D, and wall thickness, e, as well as the bulk modulus, K, defining the flowing fluid, namely

$$\frac{c}{c_0} = \sqrt{\frac{1}{1 + \dfrac{DK}{Ee}}} \tag{2.1}$$

Equation 2.1 implies that the wave propagation speed will always be less than that in the unrestricted fluid, the difference being dependent on overall pipe elasticity and fluid bulk modulus.

However the seminal work on which modern waterhammer or pressure transient theory rests was due to Joukowsky (1900) who published the results of his work at the Imperial Water Works in St Petersburg.[1] Joukowsky for the first time accurately measured wave propagation velocity agreeing with Korteweg's predictions and then went further to solve the relevant equations of continuity and motion to develop the equation that now bears his name for the pressure surge, Δp, following an instantaneous change in fluid velocity, Δu, propagated at wave speed c, in a fluid of density ρ, namely

$$\Delta p = -\rho c \Delta u \tag{2.2}$$

This expression will be returned to later as it is open to misinterpretation, in particular as it only applies to a change in flow velocity completed prior to the arrival at the source of the transient of any reflection from a system boundary, which may be either a terminal boundary, such as a free discharge, or an in-system boundary, such as a pipe junction – these caveats will be developed further.

Joukowsky's work was made available through a translation by Simin (1904) that indicated that Joukowsky achieved agreement of ±15% on pressure and ±2% on wave speed – remarkable achievements in view of the instrumentation employed. While Joukowsky is credited with these breakthroughs there has been some evidence presented by Wood (1970) that an American researcher, Frizell, developed similar expressions linking pressure to wave speed and instantaneous changes in flow velocity at the turn of the 19th century.

Joukowsky also introduced the fundamental concept of the pipe period – namely the time taken, t_p, for a transient, travelling at acoustic velocity, c, generated by a change in flow conditions to reach a system boundary and return to its source, a distance $2L$, defined as

$$t_P = \frac{2L}{c} \tag{2.3}$$

This measure is more helpful than absolute time as it determines the limit of application of Joukowsky's 'instantaneous' pressure surge equation. It will be appreciated that the time for the valve closure on a long-distance oil pipeline to remain within one pipe period will be considerably greater than that appropriate for a tank isolation valve within an in-flight refuelling system, where pipes are metres in length rather than kilometres.

Thus pressure change is dependent upon fluid density, imposed velocity change and wave speed, that, for a liquid system, itself depends on fluid properties and conduit elasticity and deformation. The final pressure regime imposed at any point in the system also depends upon the propagation wave speed and geometry of the network as, at any one location, the resultant pressure may be determined by the principle of superposition of pressure waves, whose arrival time is determined by the wave speed and system geometry, the latter determining the applicable pipe periods.

Joukowsky understood and demonstrated the pressure surge consequent upon changes in flow condition, and in particular defined the surge expected following a 'rapid' change in flow conditions, i.e. one completed in less than one pipe period. However in practice valve closures or other flow control actions such as changes in pump operating point, are likely to be 'slow', i.e. completed in more than one pipe period. Allievi (1904) developed an initial graphical methodology to allow the application of the principle of superposition for a frictionless system subjected to pressure surges generated by 'slow' changes in flow condition, and subsequently during the 1930s Schnyder (1929) and Bergeron (1932, 1961) developed more refined graphical methods that allowed the inclusion of friction. The final presentation of graphical techniques for surge analysis is due to Pickford (1969), however this work was already superseded by computer-based techniques that were introduced during the 1960s, initially as computerised versions of the graphical methods then prevalent, Harding (1966), and later as finite difference schemes based on the Method of Characteristics (MoC), that revolutionised the study of pressure surge in all of the fluid systems identified above.

The introduction of digital computing within research and industry in the early 1960s allowed pressure surge analysis to flourish. The fundamental numerical solution of the wave equations defining pressure transient propagation proposed by Lister (1960) was based on the application of the method of characteristics to transform the quasi-linear hyperbolic partial differential equations of motion and continuity into a pair of total differential equations capable of numerical finite difference solution. This technique was first utilised by Streeter et al. (1962) and his co-authors in the USA and by a range of UK authors notably led by Fox (1968). By the early 1970s regular specialist Pressure Surge Conferences (1972 to 2008) were organised by BHRA in the UK, conferences that continue to the present, attracting international contributors who presented work on all the areas already mentioned, including UK work by Boldy (1976) and Boldy and Logan (2008), Enever (1972), Thorley and Twyman (1976) and Swaffield (1972b).

The method of characteristics has become the industry standard for pressure transient analysis and applies equally to the traditional pressure surge applications as well as to the specialist area of building drainage and vent system design and analysis. It will be seen later that application of this numerical technique will allow the impact of pressure transient propagation on the operation and safety of drainage systems to be assessed and will also facilitate the development of air pressure control strategies and transient control devices to maintain system operation and minimise public health hazards arising from cross-contamination and insufficient venting.

In order to prepare for the introduction of these numerical techniques based on the Method of Characteristics solution to the wave equations, it is first helpful to both retrace the fundamental work of Joukowsky, developing his Equation 2.2, and the treatment of system boundary conditions inherent in the later work of Allievi that allowed the principle of superposition to be used to determine the pressure time history at any point in a network.

2.2   Building drainage and vent system pressure transient propagation

The air pressure regime within a building drainage and vent system is entirely dependent upon the history of air pressure transients generated by successive changes in the operating conditions of the network. Any change in the flow carried by the system must be accompanied by the propagation of a low amplitude air pressure transient, otherwise the system will be unaware of the change in operating conditions.

Consider the flow of an appliance discharge into a vertical stack leading to a sewer connection. As the flow discharges to the stack it assumes an annular flow adhering to the internal surface of the stack. As already discussed in Chapter 1, this annular flow may be several mm in thickness and will attain a terminal velocity dependent upon the water flow rate and the roughness and diameter of the vertical stack.

The air core within the stack is thus subjected locally to a downwards shear force that entrains an airflow. The air in contact with the water annulus will adopt its surface velocity under the condition of 'no slip' that applies at the interface. If the annular water downflow increases due to an additional branch appliance discharge then the increased entrained airflow demand must be communicated to the system terminations and thus a mechanism must exist to ensure that external air is drawn into the stack in sufficient quantities to maintain the airflow demanded.

The only mechanism to achieve this communication is the propagation of a low amplitude air pressure transient, negative if the air demand rate is increased, that will propagate throughout the fluid network, eventually reaching an open vent or inwards relief valve. In order to induce an airflow down the stack this transient will represent a drop in local pressure, effectively a negative air pressure wave.

Figure 2.1 illustrates this mechanism. The negative pressure wave propagates at the local acoustic velocity; in the case of building drainage systems this may be taken as the air acoustic velocity as the pipework may be considered rigid.

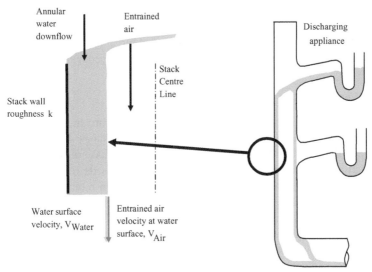

No-slip condition, $V_{Water} = V_{Air}$ generates entrained airflow

Figure 2.1 Illustration of the propagation of a transient by a change in annular water downflow communicated to the entrained air core through the shear force generated by the no slip condition

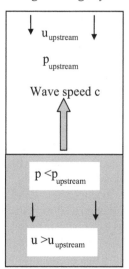

Figure 2.2 Propagation of a transient at local acoustic velocity and the associated change in flow velocity

A negative air pressure wave will induce an enhanced airflow from the upstream higher pressure zone into the low pressure zone behind the transient. The value of the induced airflow velocity will depend upon the pressure differential across the wave. Thus while the wave propagates at the acoustic velocity in air, around 340 m/s, the induced airflow has a much lower velocity, as illustrated in Figure 2.2. A suitable analogy by Gormley (2007b) is to consider the speed of transmission of a traffic hold up, communicated at the speed of recognition of traffic tail-lights compared with the speed of the traffic itself; one is independent of the other.

2.3   Wave speed

Wave speed is the most important parameter in pressure transient theory and analysis. As already shown it determines the pressure change that accompanies a change in the local fluid velocity, either positive in the case of fluid deceleration or negative as the fluid is accelerated. As important is its role in determining pipe period – the earliest time at which a reflection can return to the source of the transient from a system boundary, either internal or terminal. This property determines whether any transient is regarded as 'rapid' or 'slow' and therefore whether the full Joukowsky pressure change may be expected.

Wave propagation velocity depends on both the properties of the flowing fluid and the enclosing conduit material. In general pressure surge theory, any free gas within the liquid stream will also be a factor that will affect the calculated wave speed.

In building drainage and vent system analysis, the fluid carrying the low amplitude pressure transients is the entrained airflow generated by water discharges to the network and this introduces a number of simplifications. Generally it may be assumed that the drainage pipe and stacks are rigid when subjected to the low amplitude air pressure transients to be considered. However there may be occasions when the introduction of a highly flexible conduit would be beneficial in controlling transients and so the following derivation will demonstrate the inclusion of wall flexibility.

Wave speed or acoustic velocity in an unconfined fluid is defined as

$$c_0 = \sqrt{\frac{K}{\rho}} \qquad (2.4)$$

where K is the fluid bulk modulus, defined as (change in fluid pressure / densimetric strain (or volumetric strain)), so that $K = (\Delta p / (\Delta \rho / \rho))$, and ρ is the fluid density. For gas at normal temperatures subjected to low amplitude transients the bulk modulus, K_{gas}, may be approximated by the product of the gas absolute pressure and the ratio of specific heats, γ, so that wave speed becomes

$$c_0 = \sqrt{\frac{\gamma p}{\rho}} \qquad (2.5)$$

and

$$\frac{dp}{d\rho} = \frac{\gamma p}{\rho} \qquad (2.6)$$

However in a confined conduit the distortion of the pipe wall may also affect the wave speed. Consider the propagation of a transient through a fluid within a pipe of diameter D, wall thickness, e, Young's modulus of elasticity, E. The passage of a positive transient will compress the fluid and distend the pipe walls. The overall volumetric change for the system may be determined by the summation of these two effects.

For the gas the volumetric change becomes

$$dVol_{gas} = (Vol_{gas} / K_{gas})dp \qquad (2.7)$$

For the circular cross-section pipe the original volume of a section of pipe Δx in length is

$$Vol_{pipe} = \frac{\pi D^2 \Delta x}{4}$$

and the distended extra volume is

$$dVol_{pipe} = \pi D(\frac{\delta D}{2})\Delta x$$

For a thin walled pipe – defined as a case where $e \ll D$ – the circumferential stress, F_C, and strain, $\delta D / D$, may be written as

$$F_C = \frac{D}{2e}dp \quad \text{and} \quad \frac{\delta D}{D} = \frac{F_C}{E}$$

when longitudinal stress and strain are omitted as the pipe may be considered capable of longitudinal expansion. The volumetric strain for a pipe section of length Δx therefore becomes

$$\frac{dVol_{pipe}}{Vol_{pipe}} = \frac{dA}{A} = \frac{2\delta D}{D} = \frac{2F_C}{E} = \frac{D}{Ee}dp \tag{2.8}$$

The total change in volume as the gas is compressed and surrounding pipe distends is therefore

$$dVol_{total} = dVol_{pipe} - dVol_{gas} = \left[\frac{D}{Ee} + \frac{1}{K_{gas}}\right]dp = \left[\frac{D}{Ee} + \frac{1}{\gamma p}\right]dp$$

Thus from the definition of bulk modulus an overall effective bulk modulus may be expressed as

$$K_{eff} = \frac{1}{\left[\frac{1}{\gamma p} + \frac{D}{Ee}\right]} = \frac{\gamma p}{\left[1 + \frac{\gamma pD}{Ee}\right]} \tag{2.9}$$

and wave speed becomes

$$c = \sqrt{\frac{\gamma p}{\rho\left[1 + \frac{\gamma pD}{Ee}\right]}} \tag{2.10}$$

If the pipes carrying the fluid flow and the transients are rigid then the E term tends to infinity and the wave speed in the fluid reverts to that in an unconfined expanse of the fluid. Generally in the analysis of low amplitude air pressure transient propagation it is sufficient to assume rigid pipes so that the wave speed

becomes $c = \sqrt{\dfrac{\gamma p}{\rho}}$.

It will also be appreciated that the pressure term, p, in the wave speed must be the absolute gas pressure so that transients measured in mm of water gauge with maximum values less than 200 will have little or no effect on the calculated transient propagation velocity.

2.4   Development of the equations of continuity and momentum, leading to the Joukowsky equation

While the later finite difference based analysis of pressure transient propagation will allow the inclusion of frictional effects it is instructive to develop the equations of continuity and motion for a frictionless flow as a means of understanding the early analysis of pressure surge, the development of the Joukowsky equation and the early pressure prediction techniques based on the principle of superposition of pressure transients.

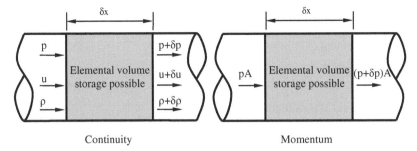

Figure 2.3 Development of the continuity and momentum equations for a one dimensional frictionless flow

Consider the one-dimensional frictionless flow illustrated by Figure 2.3. The continuity equation may be expressed as the sum of the inflow and outflows that equate to the rate of storage within the volume, hence

$$\rho A(u - (u + \delta u)) = \frac{\partial}{\partial t}(\rho A \delta x)$$

Neglecting second order terms, and dividing by $\rho A \delta x$, reduces the expression to

$$\frac{\partial u}{\partial x} + \frac{1}{\rho}\frac{d\rho}{dt} + \frac{1}{A}\frac{dA}{dt} = 0 \qquad (2.11)$$

where

- Term 1 $\partial u / \partial x$ accounts for the velocity change in the flow direction,

- Term 2 $\frac{1}{\rho}\frac{d\rho}{dt}$ accounts for the fluid compressibility, so that from Equation

  2.6 it follows that $\frac{1}{\rho}\frac{d\rho}{dt} = \frac{1}{\gamma p}\frac{dp}{dt}$

- Term 3 $\frac{1}{A}\frac{dA}{dt}$ accounts for the radial change in conduit cross-section and from

  Equation 2.8 for an elemental section length δx it follows that $\frac{1}{A}\frac{dA}{dt} = \frac{D}{Ee}\frac{dp}{dt}$

Terms 2 and 3 may be combined as $\left[\frac{1}{\gamma p} + \frac{D}{Ee}\right]\frac{dp}{dt}$ and recognised as $\frac{1}{\rho c^2}\frac{dp}{dt}$

from the general wave speed Equation 2.10. The continuity Equation 2.11 therefore reduces to

$$\rho c^2 \frac{du}{dx} + \frac{dp}{dt} = 0 \qquad (2.12)$$

Similarly the momentum equation for a frictionless one-dimensional flow, Figure 2.3, may be derived by considering the forces acting in the flow direction on a small element of fluid within the conduit.

$$\text{Force} = \text{mass} \times \text{acceleration, hence } pA - (p + dp) A = \rho A dx \frac{du}{dt}$$

so that the momentum equation becomes

$$\frac{dp}{dx} + \rho \frac{du}{dt} = 0 \tag{2.13}$$

The St Venant equations of momentum and continuity for a one-dimensional, horizontal, frictionless system may therefore be taken as

$$\frac{dp}{dx} + \rho \frac{du}{dt} = 0 \tag{2.14}$$

$$\frac{dp}{dt} + \rho c^2 \frac{du}{dx} = 0 \tag{2.15}$$

The general solution of these differential equations is due to D'Alembert and may be expressed as

$$p - p_0 = F(t + x/c) + f(t - x/c) \tag{2.16}$$

and

$$u - u_0 = -\frac{1}{\rho c} \left[F(t + x/c) - f(t - x/c) \right] \tag{2.17}$$

The $F()$ and $f()$ functions are arbitrary and may be selected to represent transients propagated through the system at the local wave speed c as a result of any change in the system operating conditions. Historically the $F()$ function is taken to represent a transient propagated upstream and the $f()$ a transient propagated downstream. This representation of pressure transient propagation was due to Joukowsky and allows the magnitude of the generated transient as a result of any change in operating conditions to be determined, as well as allowing the interaction of pressure transients with system boundary conditions to be investigated.

2.5   Transients generated by changes in flow velocity

Joukowsky was able to use equations 2.16 and 2.17 to develop the expression for the pressure rise following an instantaneous change in flow conditions that bears his name.

If the flow is controlled so that the change $(u - u_0)$ is instantaneous then, as the flow may be assumed steady prior to the instantaneous change in condition, it follows that no $f()$ function transient can be present as this would have had to have been generated by some upstream event.

Hence Equations 2.16 and 2.17 reduce to $p - p_0 = F(t + x/c)$ and $u - u_0 = -\dfrac{1}{\rho c}[F(t + x/c)]$ so that the pressure change accompanying an instantaneous control of flow is given by

$$\Delta p = p - p_0 = -\rho c(u - u_0) \tag{2.18}$$

Thus a flow brought instantaneously to rest, i.e. a velocity change of $-u_0$, will generate a pressure transient

$$\Delta p = \rho c u_0 \tag{2.19}$$

This is known as the Joukowsky waterhammer equation and can be extremely misleading. Instantaneous changes in flow condition are generally rare. In cases where the time taken to complete the change in flow condition exceeds the time taken for the propagated transient to reach a system boundary and for its reflection to return to the location of the flow change, or the point of generation of the transient, then the subsequent pressure time history will depend on the form of the returning reflection and the local pressure will deviate from the Joukowsky instantaneous pressure change. The time taken for a propagated transient to reach a reflecting boundary and return is known as the pipe period and is calculated as

$$t_P = \frac{2L}{c} \tag{2.20}$$

Thus the Joukowsky expression is true provided the flow condition change is completed in less than one pipe period. This implies that the definition of 'instantaneous' depends on the system being considered and may be a considerable period of time for long pipelines with few boundaries.

The Joukowsky equation is usually stated for the $F()$ wave – the wave transmitted upstream. However the development of the equation is equally valid for the $f()$ wave – the wave transmitted downstream if the flow condition changes. However note the sign change – a sudden cessation of flow will lead to a pressure drop on the downstream side of a control valve, Figure 2.4.

2.6 Terminal system boundary condition reflections

The reflection of incoming transients at a system terminal boundary may be determined from Equations 2.16 and 2.17.

Consider first a closed upstream boundary – effectively a dead end where the flow velocity will be zero at all times. From Equation 2.17 it follows that the incoming $F()$ wave will be reflected as an equal $f()$ wave as both u and u_0 are zero, hence $F(t+x/c) = f(t-x/c)$ and the reflection coefficient, defined as the outgoing transient / incoming transient will be

$$C_R = \frac{f(t - x/c)}{F(t + x/c)} = 1.0 \tag{2.21}$$

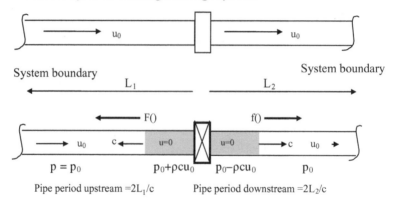

Figure 2.4 Definition of pipe period and application of the Joukowsky 'waterhammer' equation

At an open termination a similar argument may be used, together with Equation 2.16, to show that as the pressure at the open boundary is constant at its initial value, p_0, at all times it follows that

$$f(t-x/c) = -F(t+x/c)$$

so that the reflection coefficient becomes

$$C_R = \frac{-f(t-x/c)}{F(t+x/c)} = -1.0 \qquad (2.22)$$

It will be seen that the arrival of such reflections at the point of inception of the transient before the flow change is completed will affect the final pressure change and invalidates the Joukowsky expression.

Figure 2.5 illustrates a laboratory demonstration of the effect of an open or closed termination on the pressure recorded close to the point of inception of the transient. It will be seen that the sequence of positive and negative reflections from the pipe termination is changed dependent upon the imposed boundary condition. Natural attenuation of the pressure transient is also demonstrated by this dataset. It will be seen that the open-ended termination generates alternating +ve / −ve reflected pressure waves while the dead-ended system displays only positive reflected pressure waves.

2.7   Internal system boundary condition reflections

Transient reflection and transmission will also occur at any internal system boundaries, for example pipe junctions or changes in diameter. The frictionless model developed above allows the determination of the reflection and transmission coefficients applicable at each such junction.

Consider the general junction of *n* pipes illustrated in Figure 2.6. Each pipe has its individual diameter. As the flowing fluid is assumed to be air, the acoustic velocity in each pipe may be assumed constant, however individual wave speeds will be included in the development of the reflection and transmission coefficients to allow later reference to the effect of highly flexible conduits on the junction coefficients.

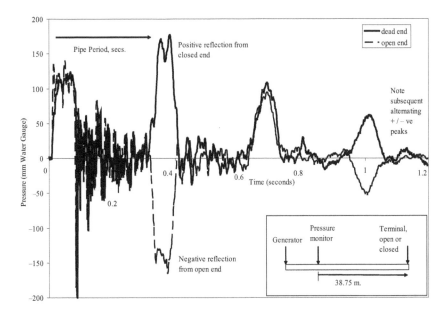

Figure 2.5 Laboratory demonstration of the +1 / -1 reflection coefficients encountered at a closed or open end termination in response to an applied pressure transient

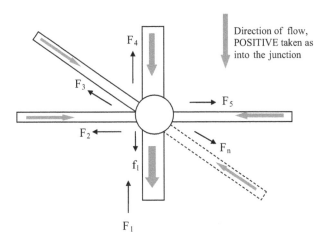

Figure 2.6 Pressure transient reflection and transmission at a multi-pipe junction as a result of a transient arriving along one pipe

If a transient arrives along one pipe due to some flow control action downstream then pressure waves will be transmitted into the flow in each of the joining pipes. There will also be a reflected wave propagated back along the pipe carrying the initial transient, however it may be assumed that no $f()$ reflected waves from upstream boundaries are present. In the absence of friction the continuity of flow and pressure at the junction may be used with the wave equations to determine the reflection and transmission coefficients.

From continuity of flow at the junction, assuming flow into the junction as positive, and no flow storage at the junction,

$$\sum_1^n Q = 0 \quad \text{so} \quad Q_1 = Q_2 + Q_3 + \ldots + Q_n \tag{2.23}$$

and $$\sum_1^n \Delta Q = 0 \quad \text{so} \quad \Delta Q_1 = \Delta Q_2 + \Delta Q_3 + \ldots + \Delta Q_n \tag{2.24}$$

In the absence of friction or any junction separation losses it also follows that

$$p_1 = p_2 = p_3 = \ldots = p_n \quad \text{and} \quad \Delta p_1 = \Delta p_2 = \Delta p_3 = \ldots = \Delta p_n$$

From the wave Equations 2.16 and 2.17, assuming only pipe 1 carries both an incoming and reflected wave

$$\Delta p_1 = F_1 + f_1 \quad \Delta Q_1 = -\frac{A_1}{\rho c_1}(F_1 - f_1), \qquad \Delta p_2 = F_2 \quad \Delta Q_2 = -\frac{A_2}{\rho c_2}(F_2),$$

$$\Delta p_3 = F_3 \quad \Delta Q_3 = -\frac{A_3}{\rho c_3}(F_3), \qquad \ldots \qquad \Delta p_n = F_n \quad \Delta Q_n = -\frac{A_n}{\rho c_n}(F_n)$$

thus from Equation 2.16

$$F_1 + f_1 = F_2 = F_3 = \ldots\ldots = F_n$$

and from Equation 2.17

$$F_1 - f_1 = \frac{c_1}{A_1}\left[\frac{A_2}{c_2}F_2 + \frac{A_3}{c_3}F_3 + \ldots.. + \frac{A_n}{c_n}F_n\right]$$

If the reflection coefficient appropriate to pipe 1 is defined as $C_{R1} = \dfrac{f_1}{F_1}$ and the transmission coefficient into pipes $i = 2$ to n from pipe 1 as $C_{Ti} = \dfrac{F_i}{F_1}$ then for the multi-pipe junction in Figure 2.6 the coefficients are

$$C_{R1} = \frac{\dfrac{A_1}{c_1} - \sum\limits_{j=2}^{j=n}\dfrac{A_j}{c_j}}{\dfrac{A_1}{c_1} + \sum\limits_{j=2}^{j=n}\dfrac{A_j}{c_j}} \tag{2.25}$$

and

$$C_{Ti} = \frac{\dfrac{2A_1}{c_1}}{\dfrac{A_1}{c_1} + \sum\limits_{j=2}^{j=n} \dfrac{A_j}{c_j}} \qquad (2.26)$$

If the junction considered is a simple change in pipe diameter as shown in Figure 2.7 then the reflection and transmission coefficients reduce to

$$C_{R1} = \frac{\dfrac{A_1}{c_1} - \dfrac{A_2}{c_2}}{\dfrac{A_1}{c_1} + \dfrac{A_2}{c_2}} \quad \text{and} \quad C_{Ti} = \frac{\dfrac{2A_1}{c_1}}{\dfrac{A_1}{c_1} + \dfrac{A_{2_1}}{c_2}} \qquad (2.27)$$

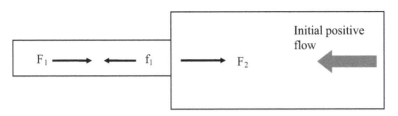

Figure 2.7 is shown with labels F_1, f_1, F_2 and "Initial positive flow".

Figure 2.7 Reflection and transmission at a change of cross section

2.8 Simplification due to constant wave speed

In a building drainage and vent system the wave propagation velocity may be taken as constant at the acoustic velocity in air as the normal piping may be considered rigid under the influence of the low amplitude air pressure transients encountered, rarely above 100 mm water gauge. Thus the expressions for transient reflection and transmission coefficients may be simplified to effective ratios of area as shown for a multi-pipe junction

$$C_{R1} = \frac{A_1 - \sum\limits_{j=2}^{j=n} A_j}{\sum\limits_{j=1}^{j=n} A_j} \qquad C_{T1} = \frac{2A_1}{\sum\limits_{j=1}^{j=n} A_j} \qquad (2.28)$$

and for a two-pipe change of section

$$C_{R1} = \frac{A_1 - A_2}{A_1 + A_2} \qquad C_{T1} = \frac{2A_1}{A_1 + A_2} \qquad (2.29)$$

The derivation of these relationships has assumed that pipe 1 carries the incoming transient. Clearly the derivation is general and if any other pipe carries the transient the expressions apply provided the carrier pipe is treated as the index pipe 1 above.

2.9   Application to a building drainage and vent system vertical stack

It will be seen from Equations 2.25 and 2.26 that generally at a multi-pipe junction the reflection coefficient will be negative and that the transient propagated into each of the upstream pipes has the same magnitude. This latter result is very important and determines the efficiency of any passive parallel venting system.

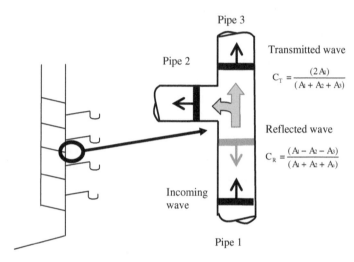

Figure 2.8 Typical three-pipe junction in a building drainage and vent system vertical stack; assume transient arrives from below along pipe 1, possibly due to a change in flow conditions further down the building

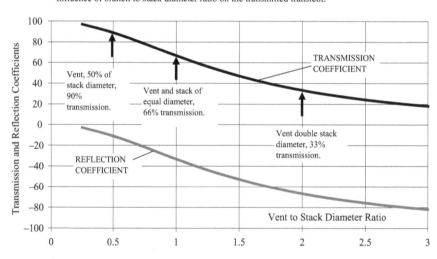

Figure 2.9 Reflection and transmission at a three-pipe junction expressed as % of the incoming wave, taken here as a positive low amplitude air pressure transient

Figure 2.8 illustrates a typical three-pipe junction while Figure 2.9 presents the reflection and transmission coefficient values for any stack to vent cross-sectional area ratio.

It will be appreciated that for the junction connection to provide any meaningful reduction in the transmitted transient, its total cross-sectional area must be considerably greater than that represented by the pipe carrying the incoming transient or its upstream continuation.

In the case of a three-pipe junction, equal diameter pipes will transmit two-thirds of the incoming transient, the remainder being reflected as a transient with change of sign. (It is best to avoid identifying the coefficients as positive or negative here as the sign of the reflected and transmitted transients depends on the sign of the incoming transient.)

As shown in Figure 2.9 the diameter of the side connection, pipe 2, would have to be double that of pipes 1 and 3 for the transmission coefficient to fall to a third of the incoming wave amplitude.

2.10 Effect of variable wave speed on junction reflection and transmission coefficients

Equations 2.25 and 2.26 define the reflection and transmission coefficients for an incident wave in one of the pipes joining at a multi-pipe junction. It will be seen that the coefficients depend upon both duct cross-sectional area and the wave speed in the duct. Normally in pressure surge research these wave speeds may vary due to changes in pipe material or pipe wall thickness, Wylie and Streeter (1978), Fox (1989). For a two-pipe junction of two equal diameter pipes it will be appreciated that there will be a reflection and transmission coefficient if the two pipes are of differing material – i.e. having differing elasticity, or pipe wall thickness – hence distorting to varying extent as a result of the pressure surge.

The application of transient theory to building drainage and vent systems is significantly simplified by the assumption that the wave speed is a constant corresponding to the acoustic velocity in air and that the pipes may be considered rigid as the pressure transients are of low amplitude – measured in mm of water gauge.

However there are examples where introducing a highly flexible branch connection may be used to change the transient reflection and transmission coefficients as a means of attenuating the propagated transient. This effect will be returned to later in considering the Active Control of transients and in particular in the design of positive air pressure attenuators.

Figure 2.10 illustrates a three-pipe junction where one of the receiving pipes is a highly deformable sheath with a much reduced wave propagation velocity. As the initial wave speed in the deformable sheath $c_2 \ll c_1$ it follows that initially the transmitted transient becomes very small. In addition the pressure within the deformable sheath will be dependent upon the inflow of air driven by a positive transient in pipe 1. This will constitute an almost constant low pressure zone, again reducing the transmitted transient as the reflection coefficient tends towards

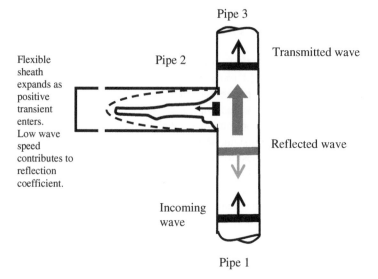

Figure 2.10 Three-pipe junction including a branch with a reduced wave speed due to the flexibility of the branch wall material

the -1 appropriate for a constant pressure terminal boundary condition. Equation 2.26 may be recast as

$$C_{Ti} = \frac{\dfrac{2A_1}{c_1}}{\dfrac{A_1}{c_1} + \dfrac{A_2}{c_2} + \dfrac{A_3}{c_3}} = \frac{\dfrac{2A_1}{c_1}}{\dfrac{2A_1}{c_1} + \dfrac{A_2}{c_2}} = \frac{1}{1 + \dfrac{A_2 / 2A_1}{c_2 / c_1}}$$

to show the effect of variable wave speed.

2.11 Trapped transient in branch

The developed expressions for transient transmission and reflection coefficients apply equally to a transient approaching a junction along any one of the connected pipes. Subsequent pressure time histories may be determined, for a frictionless system, by superposition.

Consider the pressure time history at the midpoint of a branch connecting a three-pipe junction to a system termination, either an open or closed end or an appliance trap, Figure 2.11.

The following transmission and reflection events are relevant to the subsequent pressure time history at the branch midpoint.

1　Incoming transient F1 arrives at the junction. The percentage transmitted into the branch depends upon the relative cross-sections of the pipes joining at the junction as

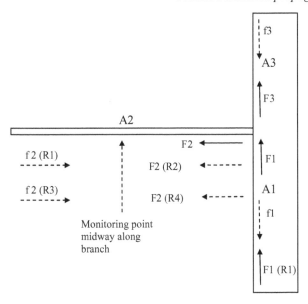

Figure 2.11 Sequence of transmitted and reflected pressure transients within a small diameter branch from a building drainage and vent system vertical stack

$$C_{T1} = \frac{2A_1}{A_1 + A_2 + A_3}$$ If $A_2 \ll A_1$ and A_3 then the transmission coefficient tends to unity.

2 Transient F2 then propagates along the branch, passing the monitoring point at time $(0.5L_2/c)$ and reaches the branch termination at time L_2/c. It is then reflected as the f2 (R1) wave that arrives at the monitoring point midway along the branch at time $(1.5L_2/c)$, continuing towards the junction, where it arrives at time $(2L_2/c)$.

3 Transient f2 (R1) is itself reflected at the junction. In this case the relevant reflection coefficient is given by

$$C_{R2} = \frac{A_2 - A_1 - A_3}{A_2 + A_1 + A_3}$$

Note that pipe 2 now carries the incoming transient and hence becomes the index pipe. Therefore if $A_2 \ll A_1$ and A_3 it follows that the reflection coefficient approaches -1.

4 Transient f2 (R1) thus becomes a reflected transient F2 (R2) as shown and this propagates towards the branch termination, passing the monitoring station at time $(0.5L_2/c)$ later and reaching the branch termination at time L_2/c later.

5 The process described above now repeats until the pressure wave attenuates due to friction – not included in this simplistic model. Reflections f2 (R3) and F2 (R4) would be generated as described above.

6   At some time in the future, reflections of the F3 transient and a re-reflection of the f1 transient, initially generated at the junction, will be generated by the system terminations of pipes A1 and A3 and will arrive at the junction and be reflected and transmitted as described above, further complicating the pressure time history at the midpoint of pipe 2.

7   The sign of the reflected waves arriving at the junction will be wholly dependent upon the system terminal boundary conditions for pipes 1 to 3. Open ends will result in a change of sign of the incoming transient at the boundary while closed ends replicate the sign and magnitude of the incoming wave. Various other boundary conditions may have variable reflection coefficients, for example an air admittance valve will act as a partially open end until the valve closes if the incoming transient is positive. Similarly an appliance trap may act as an open end if the incoming transient is sufficient to deplete the trap seal. A variable volume containment device to limit positive air pressure transient propagation will alter its reflection coefficient from an open end value to a closed end if the containment bag becomes pressurised.

2.12   Principle of superposition of pressure waves

The example above provides sufficient information for the pressure time history at the midpoint of the branch to be predicted for both open and closed branch terminations. It is helpful to use the branch pipe period, $2L_2/c$, as the unit of time. If the branch cross-sectional area is 10% of the main stack cross-section then the initial transmission from pipe 1 to pipes 2 and 3 would be 95.23%.

Similarly the reflection coefficient as the branch transient f2 approaches the junction would be 90.47% of the incoming f2 wave. This reflection coefficient < 1 eventually reduces the transient trapped in the branch, as shown by Figure 2.12.

The reflection coefficient at the branch termination will be taken as either 1 or −1 dependent upon the choice of open or closed end condition.

The pressure time history at the branch midpoint may then be determined as shown in Figure 2.12.

It should be noted that in the case of the closed end branch termination the branch pressure tends towards a trapped pressure value close to the initial transient propagated into the branch, in this case 95.23% of the initial transient in pipe 1. In the open-ended branch case the final pressure tends to atmospheric.

The demonstration of the principle of superposition in Figure 2.12 is simplistic as it has been assumed that the transients are all 'rapid', i.e. representing pressure changes accomplished either instantaneously as in the example or in less than one pipe period – in this case it is the branch pipe period which is of interest. If the case of a 'slow' pressure change is considered then the limitations of this approach will become clear.

Assume that the transient arriving at the junction along pipe 1 to initiate the process had a 'rise time', i.e. the time taken for the transient pressure level to increase to its maximum, equivalent to 1.5 times the branch pipe period. This

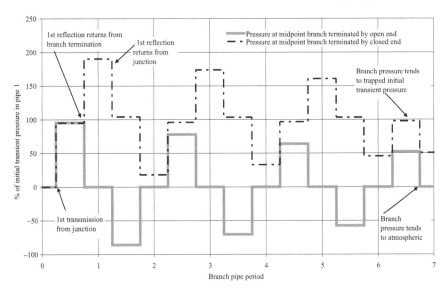

Figure 2.12 Pressure time history at the midpoint of the branch illustrated in Figure 2.11 due to the transmission and reflection of an instantaneous wave front within the branch, for both open and dead ended branch terminations

implies that as it is transmitted into the branch there is time for the leading edge of the transient to travel to the branch termination and return to the junction. Thus there are immediately two transients contributing to the pressure time history at the monitoring point. Subsequent reflections from the junction and the branch termination will also overlie each other.

The principle of superposition allows each transient to be considered in isolation and the total pressure summed at any instant in time. Figures 2.13 and 2.14 illustrate the process for both open-ended and dead-ended branches for the transients F2 and f2(R1,3,*n*–*1*) and F2(R2,4,*n*) and demonstrate the complexity of even such a simple summation.

As already mentioned, at some later time reflections from the system terminations to pipes 1 and 3 will arrive at the junction. If pipe 3 is terminated as an open-topped stack then a negative reflection will arrive at the junction and pass into the branch as before. This will further complicate the pressure time history at the midpoint of the branch and demonstrates why the graphical methods developed from Alleivi's initial work by Schnyder and Bergeron gave way so quickly to the computer-based techniques introduced in the 1960s. While it is possible to use the principle of superposition to build the pressure time history, it will be appreciated that this becomes complex if the wave fronts considered are not instantaneous – as in this simplified example. Similarly no account of friction can be made in this approach, although the Bergeron graphical method did attempt this.

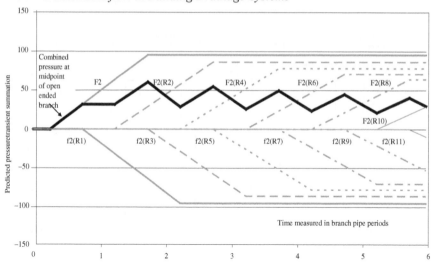

Figure 2.13 Pressure time history midway along an open ended branch subject to 'slow' rise time transients

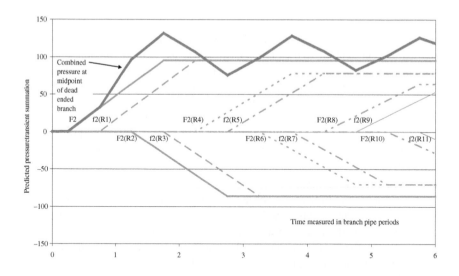

Figure 2.14 Pressure time history midway along a dead ended branch subject to 'slow' rise time transients

The advantages of a computerised analysis based on a finite difference approach will be presented in later chapters and will form the basis for the analysis methodology presented in this book.

2.13 Trap seal reflection coefficient

If a branch is terminated in an appliance trap seal then that will constitute a system boundary and will reflect incoming transients. At first sight it may be thought that the appliance trap seal water column would represent a dead-ended boundary condition, however the trap seal will be deflected in response to the incoming transient and an oscillation will result that may lead to trap seal depletion.

Consider first the likely reflection coefficient at the water seal to branch air column interface. If the water column represented a dead end the air velocity imposed at the interface would continue to be zero and the resulting reflection coefficient would be +1.

However as the trap will move in response to the applied pressure transient it is possible to suggest a continuation velocity which will result in a reflected transient of slightly lower magnitude than the incoming wave, effectively a positive reflection coefficient less than unity.

Applying the equation of motion to the trap seal column, assuming a water mass, m_w, allows a determination of the initial acceleration of the water column, Figure 2.15:

$$(p_{atmosphere} - p_{branch} + \rho g \Delta H) A_{trap} = m_w \frac{\Delta u}{\Delta t}$$

The trap movement therefore allows a continuation velocity u_t which is independent of the branch airflow velocity induced by the transient, u, so that the reflected air pressure transient becomes

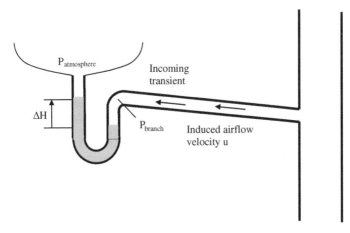

Figure 2.15 Trap seal water column subjected to incoming transients from a building drainage and vent system vertical stack

$$p = p_{branch} - \rho c(u - u_t)$$

Clearly this value is dependent upon the mass of the trap seal column and the frictional condition of the trap and will change with time as the water column is displaced and the ΔH term increases and p_{branch} responds to the reflected wave. Trap seal water may also be lost, altering the column mass. This would be impossible to represent using the superposition of pressure waves discussed earlier and is best modelled using a finite difference approach where the trap response equation becomes the boundary equation for the branch.

2.14 Concluding remarks

Low amplitude air pressure propagation within building drainage and vent systems has been introduced for a frictionless system and the roots of the necessary analysis has been developed, based on a century-long international research and application effort initiated by Joukowsky in Imperial Russia in 1900 and carried forward by authors and researchers drawn from across the whole spectrum of the international engineering and mathematical disciplines.

While the techniques introduced are essential to the understanding of transient propagation, and invaluable in the art of interpretation of system pressure response to changes in system operating condition, the limitations of the basic approaches described have also been highlighted. This leads naturally to the introduction of more sophisticated numerical analysis techniques based heavily on ground-breaking research in the 1960s that led to the identification of the Method of Characteristics as the industry standard for the solution of the St Venant wave equations. These developed techniques are applicable not only in the traditional engineering areas where waterhammer and pressure surge has long been recognised as an inevitable design condition, but also within the specialist field represented by this treatment of building drainage and vent system operation.

Note

1 Nikolai Yegorovich Zhukovskiy (Joukowsky) (January 17 [O.S. January 5] 1847 – March 17, 1921), Russian scientist and founding father of modern aero- and hydro-dynamics. Born in the village of Orekhovo, Vladimir Oblast, he graduated from Moscow University in 1868. From 1872 he was a professor at the Imperial Technical School and established the world's first Aerodynamic Institute in 1904 in Kachino near Moscow. From 1918 he was the head of TsAGI (Central AeroHydroDynamics Institute). He was the first engineer scientist to explain mathematically the origin of aerodynamic lift, through his circulation hypothesis; recognise that lift forces generated by a body moving a fluid are proportional to the velocity squared; develop a mathematical conformal transformation to define the shape of an aerodynamic profile comprising a rounded nose (leading edge), finite thickness, cambered or straight, and a sharp trailing edge. He built the first wind tunnel in Russia. He is also responsible for the eponymous waterhammer equation used by civil engineers. A city near Moscow and the Zhukovskiy (Joukowsky) crater on the Moon are named in his honour.

3 Mathematical basis for the simulation of low amplitude air pressure transients in vent systems

The mechanism by which entrained airflow is established within drainage and vent systems is well understood, Figure 3.1. The annular water flows present in the 'wet' stack entrains an airflow due to the condition of 'no slip' established between the annular water and air core surfaces. This results in the expected

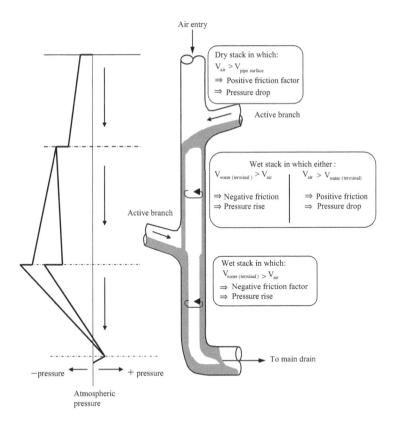

Figure 3.1 Expected pressure regime in a building drainage network vertical stack subject to multiple appliance discharges

pressure variation down a vertical stack, falling from atmospheric above the stack entry due to friction and the effects of drawing air through the vent entry and the water curtains formed at each successive discharging branch junction. Within the lower wet stack region the pressure recovers to above atmospheric due to the traction forces exerted on the airflow and the necessity to discharge air to the downstream drain through the water curtain formed at the stack base. These mechanisms may be used as a basis for a mathematical simulation of air entrainment provided that the relationships linking applied waterflow to entrained airflow are known, together with the influence of stack diameter, roughness and building height. These relationships have been identified and utilised in the development of the simulation model AIRNET, to be described in this chapter, which allows the application of the finite difference Method of Characteristics to predict the time dependent air pressure regime in the system as a result of appliance discharge.

The application of the Method of Characteristics to the modelling of unsteady flows was first recognised in the 1960s, Lister (1960), Streeter et al. (1962). The resulting solution of the equations of continuity and momentum has been applied to free surface drainage flows as well as entrained airflows and waterhammer, Swaffield and Galowin (1992), Swaffield and Campbell (1992a and b, 1995), Swaffield and Boldy (1993). The relationships defined by Jack (2000) allow the current MoC simulation to predict air transient propagation by introducing a pseudo-friction factor applicable between the annular water film and the air core, thus providing a modelling of the traction force exerted on the entrained air, Swaffield (2006).

Jack (2000) proposed that, by summing each of the defining pressures within a vertical stack system, i.e. pressure losses through each active branch, the total frictional losses and stack entry loss as well as the back pressure to be overcome at the stack base water curtain, the 'traction work done' by the falling annular water film could be calculated. This approach draws on earlier work, Lillywhite and Wise (1969), that attempted to model steady annular downflow through application of the steady flow energy equation. Based on extensive experimental data this approach allows the definition of a 'pseudo-friction factor' applicable in the wet stack and operable at the water annular flow / entrained air core interface. The most important outcome is that predictions need not be limited to single point discharge. Through the use of a variable friction factor term, combined discharge flows and their effect on the entrainment of air can be modelled. It will be appreciated that the airflow entrained in the lower levels of the wet stack may exceed that appropriate to the annular water flows present in the upper levels. The variable friction factor allows the lower level annular flow to provide airflow entrainment and hence a rising air pressure. In the upper levels this air is effectively drawn down past a slower moving water film that impedes its entrainment, and leads to an observed reduction in air core pressure levels. When linked to a Method of Characteristics solution, this approach provides a general simulation applicable across the whole range of vent system design.

3.1 Development of the general St Venant equations of continuity and momentum

The defining unsteady flow equations of continuity and momentum were developed in the mid-19th century by St Venant (1870) as a pair of quasi-linear hyperbolic partial differential equations that may be solved by numerical techniques once transformed into a pair of total derivative expressions through the use of the Method of Characteristics, first used by Riemann in 1860 to study sound wave propagation and later by Massau in 1900 to consider free surface wave motion. The first application to pressure surge analysis is due to Lamoen (1947), Gray (1953) and Ezekial and Paynter (1957), however none of these contributions made reference to the impact of access to fast digital computing. Lister (1960) established the application of MoC to solve the waterhammer equations and her application was taken up by Streeter and Lai (1962) and his co-authors at the University of Michigan. European and UK involvement was led by Fox (1968) at Leeds University and by a group of researchers at other UK universities, including Boldy, Enever, Swaffield, Thorley and Twyman, leading in 1972 to the first in the series of BHRA Pressure Surge Conferences that continue to the present day, BHRA (now BHR) (1972–2008).

Figures 3.2(a) and (b) illustrate the unsteady flow conditions in an element of one-dimensional flow between two sections fixed in space d*s* apart, such that $\partial s / \partial t = 0$, within a conduit of general cross-section that may or may not be flowing full. The conduit is assumed to be elastic and to be only subjected to small deformations due to the time dependent pressure regime within the conduit. The conduit and the enclosed fluid do not move together as a rigid body – any change in conduit section length affects the conduit cross-section through the action of

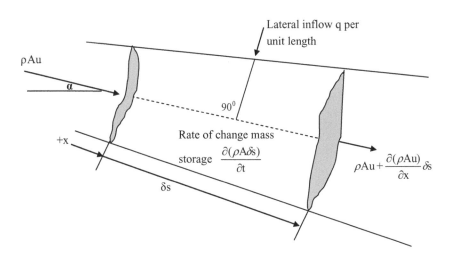

Figure 3.2(a) Derivation of the continuity equation for unsteady flow in a general conduit

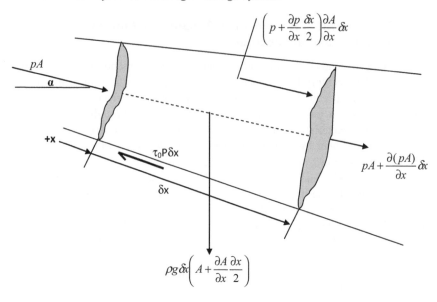

Figure 3.2(b) Derivation of the momentum equation for unsteady flow in a general conduit

the material's Poisson ratio. In order to retain a general solution the possibility of lateral inflow or extraction, q, is included in the derivation.

The continuity equation may be expressed, Figure 3.2(a), as

Mass inflow – outflow = rate of change of mass storage within the flow element

$$\rho Au - \left[\rho Au + \frac{\partial(\rho Au)\delta s}{\partial x} \right] + q\rho\delta s = \frac{\partial(\rho A\delta s)}{\partial t} \tag{3.1}$$

Expanding, collecting terms and dividing by $\rho A\delta s$ yields the continuity equation

$$\frac{\partial u}{\partial x} + \frac{1}{A}\left[\frac{\partial A}{\partial t} + u\frac{\partial A}{\partial x} \right] + \frac{1}{\rho}\left[\frac{\partial \rho}{\partial t} + u\frac{\partial \rho}{\partial x} \right] - \frac{q}{A} = 0 \tag{3.2}$$

simplified as

Term 1c + Term 2c + Term 3c – Term 4c = 0

to allow a description of the influence of each term in turn and to identify which terms are appropriate etc. in any particular unsteady flow application, Douglas et al. (2005). Term 2c represents the effect of conduit cross-sectional change while Term 3c represents the effect of density changes.

Similarly the equation of momentum may be expressed, Figure 3.2(b), as

$$pA + \left(p + \frac{\partial p}{\partial x} dx \right) \left(A + \frac{\partial A}{\partial x} dx \right) + \left(p + \frac{1}{2} \frac{\partial p}{\partial x} dx \right) \frac{\partial A}{\partial x} dx - \tau_0 P dx + mg \sin \alpha$$

$$= \rho A dx \left(\frac{\partial u}{\partial t} + \frac{\partial u}{\partial x} \frac{dx}{dt} \right) + \rho q dx u \tag{3.3}$$

It is to be noted that the lateral inflow q is included in the momentum term and that in the general case the conduit slope introduces a gravitational force. As the element is of general cross-section there is a force component derived from the pressure acting on the outer surfaces of the element. The wetted perimeter is referred to as P and the shear stress acting between the fluid and the conduit wall is τ_0.

Expanding, neglecting second order terms and dividing by ρA leads to the general momentum equation

$$\frac{1}{\rho} \frac{\partial p}{\partial x} + \left(\frac{\partial u}{\partial t} + \frac{\partial u}{\partial x} \frac{dx}{dt} \right) - g \sin \alpha + \frac{\tau_0 P}{\rho A} + \frac{qu}{A} = 0 \tag{3.4}$$

It is again convenient to represent this expression as

Term 1m + Term 2m − Term 3m + Term 4m + Term 5m = 0

Each term has its particular influence, e.g. Term 3m represents the effect of conduit slope, Term 4m the shear forces between the flow and the enclosing surface and Term 5m the lateral inflow contribution to flow momentum change.

Table 3.1 illustrates the necessity to include each of these terms in either the continuity or momentum equation dependent upon the unsteady flow regime to be considered, Douglas et al. (2005).

Table 3.1 Relevance of each term identified within the St Venant unsteady flow equations of continuity and momentum.

Unsteady flow regime	Continuity Equation 3.2	Momentum Equation 3.4
Traditional waterhammer, including closed conduit flows, siphonic rainwater systems.	Terms 1, 2, 3	Terms 1, 2, 3, 4
Free surface flows, including building drainage systems within or close to the building envelope.	Terms 1, 2 only	Terms 1, 2, 3, 4
Free surface flows where lateral inflow becomes important.	Terms 1, 2, 4 only	Terms 1, 2, 3, 4, 5
Low amplitude air pressure transient applications, simplifications include no changes in conduit cross section, no lateral inflows or longitudinal duct extensions.	Terms 1, 3 only	Terms 1, 2, 4 only

3.2   Derivation of the St Venant unsteady flow equations in the special case of low amplitude air pressure transient propagation in building drainage and vent systems

Based on the terms identified as specifically relevant to low amplitude air pressure transient propagation, the equations of continuity and momentum may be simplified to

$$\frac{\partial u}{\partial x} + \frac{1}{\rho}\left[\frac{\partial \rho}{\partial t} + u\frac{\partial \rho}{\partial x}\right] = 0 \qquad (3.5)$$

$$\frac{1}{\rho}\frac{\partial p}{\partial x} + \left(\frac{\partial u}{\partial t} + \frac{\partial u}{\partial x}\frac{dx}{dt}\right) + \frac{\tau_0 P}{\rho A} = 0 \qquad (3.6)$$

It will be seen that Equations 3.5 and 3.6 are cast in terms of pressure, p, density, ρ, and velocity, u, as well as distance, x, and time, t, however gas pressure and density are linked if it is assumed that the fluid is a perfect gas and that the flow is isentropic. These are acceptable assumptions in building drainage and vent system applications as the changes in temperature and pressure are small during transient propagation so that the process may be considered reversible. As the temperature gradients are small, and the process is rapid, no heat is transferred, hence the process is considered isentropic and air pressure and density may be linked throughout the flow as

$$\frac{p}{\rho^\gamma} = k \qquad (3.7)$$

where k is a constant.

It was shown in Chapter 2 that wave speed or acoustic velocity in an unconfined fluid is defined as

$$c = \sqrt{\frac{K}{\rho}} \qquad (3.8)$$

where K is the fluid bulk modulus, defined as (change in fluid pressure / densimetric strain (or volumetric strain)), so that $K = \left(\Delta p / (\Delta \rho / \rho)\right)$, and ρ is the fluid density. For gas at normal temperatures subjected to low amplitude transients the bulk modulus, K_{gas} may be approximated by the product of the gas absolute pressure and the ratio of specific heats, g, so that wave speed becomes

$$c = \sqrt{\frac{\gamma p}{\rho}} \qquad (3.9)$$

In order to utilise Equations 3.5 and 3.6 it is necessary to first define the relationship between pressure and density in order to reduce the variables. Consider the propagation of a transient through a one-dimensional gas flow, Figure 3.3. If the transient is brought to rest by the superposition of a velocity $-c$ onto the system it is possible, by reviewing the equations of continuity and

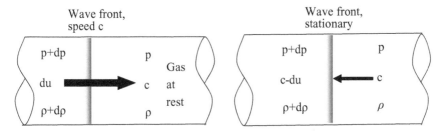

Figure 3.3 Low amplitude air pressure transient in a drain or vent, cross-sectional area A, brought to rest by the superposition of a –c air velocity

momentum, to develop a relationship linking pressure, p, and density, ρ, to wave speed, c.

From Figure 3.3, if shear forces are neglected then the momentum equation becomes

$$(p+dp)A - pA = cA\rho\left[-(c-du)-(c)\right] \qquad (3.10)$$

so that $dp = \rho c du$. $\qquad (3.11)$

Similarly the continuity equation, with the transient brought to rest by the superposition of an equal but opposite system velocity, becomes

$$cA\rho = (c - du)A(\rho + d\rho) \qquad (3.12)$$

so that $d\rho = \dfrac{\rho}{c}du$ $\qquad (3.13)$

and hence $\dfrac{dp}{\rho c} = \dfrac{c}{\rho}d\rho$ $\qquad (3.14)$

and from Equations 2.5, 3.7 and 3.14

$$c = \sqrt{\frac{dp}{d\rho}} = \sqrt{k\rho^{\gamma-1}} = \sqrt{\gamma\frac{p}{\rho}} \qquad (3.15)$$

These wave speed and pressure/density relationships are strictly only applicable to inviscid flows, however previous work, particularly in the area of train/tunnel interactions, Henson and Fox (1974a, 1974b) and Woodhead, Fox and Vardy (1976), have indicated close agreement between experimental pressure responses and transient predictions utilising these simplifications. This approach is particularly suited to the current application where the ambient pressure and the transient pressure excursion are both low. In building drainage system analysis the ambient pressure is atmospheric and the transient is not likely to exceed 100 mm of water gauge. In cases outwith these conditions Vardy (1976) has introduced the additional energy equation to allow the analysis to proceed, however that is not necessary in this application or others where the low transient pressure is maintained, Wiggert and Martin (2008).

From Equation 3.15

$$c = k^{\frac{1}{2}}\rho^{\frac{\gamma-1}{2}} \text{ and } \frac{\partial c}{\partial \rho} = \frac{\gamma-1}{2}k^{\frac{1}{2}}\rho^{\frac{\gamma-1}{2}-1} = \frac{\gamma-1}{2}\frac{c}{\rho^{\frac{\gamma-1}{2}}}\rho^{\frac{\gamma-1}{2}-1} = \frac{\gamma-1}{2}c\rho^{-1}$$

hence

$$\partial\rho = \left[\frac{2}{\gamma-1}\right]\frac{\rho}{c}\partial c \tag{3.16}$$

and similarly

$$\partial p = \left[\frac{2}{\gamma-1}\right]\rho c \partial c \tag{3.17}$$

Equations 3.16 and 3.17 allow pressure and density to be replaced in the St Venant equations by the wave speed, c, reducing these expressions to two dependent and two independent variables, wave speed and flow velocity and time and distance. The equations of continuity and momentum in c and u therefore become

$$L_1 = \rho\frac{\partial u}{\partial x} + \left[\frac{2}{\gamma-1}\right]\frac{\rho}{c}\left[\frac{\partial c}{\partial t} + u\frac{\partial c}{\partial x}\right] = 0 \tag{3.18}$$

$$L_2 = \left[\frac{2}{\gamma-1}\right]\frac{\rho}{c}\frac{\partial c}{\partial x} + \rho\left(\frac{\partial u}{\partial t} + u\frac{\partial u}{\partial x}\right) + \frac{4\rho fu|u|}{2D} = 0 \tag{3.19}$$

when the shear stress term is replaced by $\tau_0 = f\frac{1}{2}\rho u^2$, where f is friction factor and the velocity squared term is replaced by $u|u|$ to ensure that the sign of the friction term always opposes motion in a dry stack – note the appropriate representation of shear stress in the wet stack where the entrained airflow is in contact with the water annular downflow will be returned to later – and the hydraulic mean depth term, A/P, is replaced by its value for a circular section duct, namely $D/4$.

3.3   Application of the Method of Characteristics to transform the St Venant equations into a total derivative form

While Equations 3.18 and 3.19 represent the continuity and momentum relationship in an unsteady airflow within a building drainage and vent system, in the form presented they are not amenable to numerical solution. However Lister (1960) presented a generalised approach, the Method of Characteristics (MoC), that allows this pair of quasi-linear hyperbolic differential equations in two dependent and two independent variables to be transformed into a pair of total differential equations that may be solved numerically.

Equations 3.18 and 3.19 may be combined as

$$L = \lambda L_1 + L_2 = 0$$

where λ is an arbitrary term chosen to allow the transformation of these differential expressions, hence

$$L = \frac{\partial u}{\partial t} + (\lambda + u)\frac{\partial u}{\partial x} + \left(\frac{2}{\gamma - 1}\right)\left[\frac{\lambda}{c}\frac{\partial c}{\partial t} + \left(\lambda\frac{u}{c} + c\right)\frac{\partial c}{\partial x}\right] + \frac{4\,fu|u|}{2D} = 0 \qquad (3.20)$$

$$L = \frac{\partial u}{\partial t} + (\lambda + u)\frac{\partial u}{\partial x} + \left(\frac{2}{\gamma - 1}\right)\frac{\lambda}{c}\left[\frac{\partial c}{\partial t} + \left(u + \frac{c^2}{\lambda}\right)\frac{\partial c}{\partial x}\right] + \frac{4\,fu|u|}{2D} = 0$$

As $u = \theta(x,t)$ and $c = \theta(x,t)$ it follows that

$$\frac{du}{dt} = \frac{\partial u}{\partial t} + \frac{\partial u}{\partial x}\frac{dx}{dt} \text{ and } \frac{dc}{dt} = \frac{\partial c}{\partial t} + \frac{\partial c}{\partial x}\frac{dx}{dt}$$

so that $\dfrac{dx}{dt} = (\lambda + u) = (u + \dfrac{c^2}{\lambda})$ and hence $\lambda = \pm c$.

Equation 3.20 thus reduces to a total differential equation

$$\frac{du}{dt} \pm \frac{2}{\gamma - 1}\frac{dc}{dt} + \frac{4\,fu|u|}{2D} = 0 \qquad (3.21)$$

provided that

$$\frac{dx}{dt} = u \pm c \qquad (3.22)$$

The form of the combined Equation 3.21 is suitable for finite difference solution, however the dx/dt expressions represented by Equation 3.22 place restrictions on the solution and give the method its title as Equation 3.22 defines the slope of the 'characteristics' that may be drawn in an x–t plane to represent the characteristic equations 3.21. Figure 3.4 illustrates the solution technique as finite difference expressions are constructed to represent the relationship between known conditions at points R and S and future unknown values of u and c at point P, one time step into the future. It will be appreciated that as the slopes of RP and SP are slightly different – i.e. $dx/dt = u \pm c$, the characteristics do not pass through A and B unless u is negligible compared to c.

Thus the C^+ characteristics linking RP may be expressed as

$$u_P - u_R + \frac{2}{\gamma - 1}(c_P - c_R) + 4 f_R u_R \,|\, u_R \,|\,\frac{\Delta t}{2D} = 0 \qquad (3.23)$$

when

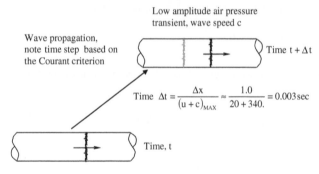

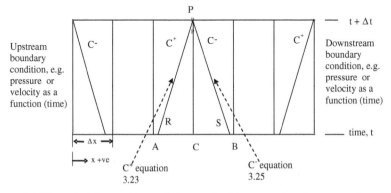

Figure 3.4 Characteristic equations within an x–t grid representing the propagation of low amplitude air pressure transients in a building drainage and vent system

$$\frac{dx}{dt} = u_R + c_R \tag{3.24}$$

and the C^- characteristics linking SP may be expressed as

$$u_P - u_S - \frac{2}{\gamma - 1}(c_P - c_S) + 4f_s u_S \,|u_S|\,\frac{\Delta t}{2D} = 0 \tag{3.25}$$

when

$$\frac{dx}{dt} = u_S - c_S \tag{3.26}$$

where the wave speed c, in terms of a base condition, normally atmospheric pressure, $p_{0,}$ and density, ρ_0, is given by

$$c = \sqrt{\frac{\gamma p}{\rho}} = \sqrt{\frac{\gamma p}{\left[\dfrac{\rho \rho_0^\gamma}{p_0}\right]^{\frac{1}{\gamma}}}} = \sqrt{\frac{\gamma p^{\frac{\gamma-1}{\gamma}}}{\dfrac{\rho_0}{p_0^{\frac{1}{\gamma}}}}} = \sqrt{\gamma p^{\frac{\gamma-1}{\gamma}}\,\frac{p_0^{\frac{1}{\gamma}}}{\rho_0}} \tag{3.27}$$

The choice of u and c as the solved variables in equations 3.23 and 3.25 resulted from the interdependence of air pressure and density, therefore it is necessary to determine pressure at each node at each time step from the equation as

$$p_{local} = \left[(\frac{p_0}{\rho_0^\gamma})(\frac{\gamma}{c_{local}^2})^\gamma \right]^{\frac{1}{1-\gamma}}$$ (3.28)

It has been stated earlier that in this particular application it is acceptable to assume that wave speed will be constant at the acoustic velocity in air. Clearly as the characteristic equations are solved for flow velocity and wave speed this is not wholly accurate and the resulting pressure determined by Equation 3.28 depends upon the variability of the wave speed. However the influence of pressure on wave speed within the likely range applied in building drainage and vent system analysis is small, 0.5 per cent for a pressure variation of ± 200 mm water gauge, Kelly (2009b). Thus it is acceptable to treat wave speed in this application as sensibly constant while still recognising its role in the solution of the characteristics equations.

Figure 3.4 illustrates the Courant criterion, represented by equations 3.24 and 3.26 that link internodal distance, Δx, to time step, Δt, in terms of the airflow velocity, u, and wave speed, c. In practice it is necessary for the same time step to be used in all calculations at any one time across the whole network. Thus the appropriate time step becomes the smallest to satisfy the Courant criterion,

$$\Delta t = \frac{\Delta x}{(u + c)_{MAX}}$$ (3.29)

If a time step is imposed on the grid illustrated in Figure 3.4 then it is necessary to interpolate between known values of u and c at node points A, B, C at time t to determine the base conditions at R and S to allow solution of the C^+ and C^- equations that intersect at P.

It is necessary to develop interpolation expressions that will allow the conditions at R and S to be determined from previously calculated values of u and c at the nodal points A, B, C.

For the nodes separated by the time step it follows that $x_p = x_c$ and by geometry

$$x_P - x_R = (u_R + c_R)\Delta t .$$

Interpolation allows

$$\frac{c_C - c_R}{c_C - c_A} = \frac{u_C - u_R}{u_C - u_A} = \frac{x_C - x_R}{x_C - x_A} = (u_R + c_R)\frac{\Delta t}{\Delta x} = \frac{p_C - p_R}{p_C - p_A}$$

and elimination results in the following interpolation relationships

$$u_R = \frac{u_C + \theta(-u_C c_A + c_C u_A)}{1 + \theta(u_C - u_A + c_C - c_A)}$$ (3.30)

$$c_R = \frac{c_C\left(1-u_R\theta\right)+c_Au_R\theta}{1+c_C\theta-c_A\theta} \tag{3.31}$$

$$p_R = p_C - \theta\left(u_R + c_R\right)\left(p_C - p_A\right) \tag{3.32}$$

$$u_S = \frac{u_C - \theta\left(u_Cc_B - c_Cu_B\right)}{1-\theta\left(u_C - u_B - c_C + c_B\right)} \tag{3.33}$$

$$c_S = \frac{c_C\left(1+u_S\theta\right)-c_Bu_S\theta}{1+c_C\theta-c_B\theta} \tag{3.34}$$

$$p_S = p_C + \theta\left(u_S - c_S\right)\left(p_C - p_B\right) \tag{3.35}$$

where $\theta = \dfrac{\Delta t}{\Delta x}$.

The characteristic equations, 3.23 and 3.25, may be reduced at any time to

$$u_P = K1 - K2c_P \tag{3.36}$$

for the C^+ expression and

$$u_P = K3 + K4c_P \tag{3.37}$$

for the C^- expression where

$$K1 = u_R + \frac{2}{\gamma-1}c_R - 4f_Ru_R\left|u_R\right|\frac{\Delta t}{2D}$$

$$K3 = u_S - \frac{2}{\gamma-1}c_S - 4f_Su_S\left|u_S\right|\frac{\Delta t}{2D}$$

$$K2 = K4 = \frac{2}{\gamma-1}$$

While interpolation is necessary if the wave speed varies across the network and it is necessary to impose a minimum time step on the calculations in order to ensure that the simulation proceeds in an orderly fashion, e.g. it is necessary to know the values of the variables u, c and p at the exit of all pipes joining at a junction at the same time instant, care must be taken to avoid the rounding errors inherent in the interpolation process. In order to calculate the base conditions at R and S at time t, it is necessary to assume that transients that reached A and B at time t affect conditions at R and S at that time – this leads to a rounding of the wave front and can introduce unrealistic damping of the pressure response.

In the case of low amplitude air pressure transient analysis, the wave speed, c, remains sensibly constant throughout the transient event as the pressure variation is minute when compared with atmospheric pressure. The necessary interpolation is therefore a minimum as the points R and S will tend to A and B. Wiggert and Martin (2008) utilise the liquid as opposed to gaseous form of the St Venant equations, with pressure and flow velocity as the variables and treating density and wave speed as constants, removing the necessity for interpolation and generating a rectangular grid in the x–t plane.

3.4 Frictional representation within the St Venant equations

Shear forces resisting motion are naturally included in the momentum Equation 3.4 and therefore appear in the finite difference expressions 3.23 and 3.25. Generally the friction factor, f, in these equations is represented by a value determined from the application of the Colebrook–White equation where it is assumed that the rate of change of the flow conditions is sufficiently slow for the steady state value of the friction factor, based on the local mean flow velocity, to be sufficiently accurate. This is effectively a quasi-steady flow approach where the Colebrook–White equation is expressed as

$$\frac{1}{f} = -4\log_{10}\left(\frac{k}{3.71D} + \frac{1.26}{\mathrm{Re}\sqrt{f}}\right) \tag{3.38}$$

The pipe wall roughness ratio k/D is assumed from available data and the mean flow velocity Reynolds number is taken as $\mathrm{Re} = \rho u D / \mu = u D / \upsilon$, with the boundary between laminar and turbulent flows being accepted as the 2000 value for circular section pipework.

Generally this is an acceptable approach as the rate of change of the flow conditions is inherently linked to the rate of change of the driving functions. In this case it is the appliance discharges to the network that entrain the airflow and therefore result in transient propagation. Generally rise times, i.e. the time taken for the appliance discharge to reach a peak value, are seldom less than 2 or 3 seconds so the transients generated are generally 'slow', i.e. the transient rise time is long in terms of the network pipe periods. For example in a 20-storey building, stack height 60 m, the stack base to roof pipe period is around a third of a second – much shorter than the likely appliance discharge rise time.

The frictional representation in the characteristic Equations 3.23 and 3.25 may be described as a first order approximation as the loss is based on the previously determined velocities at the node points A, B, C and the interpolated points R and S. This allows the direct simultaneous solution of the characteristics for all internal nodes, 2 to N, in any pipelength. Early applications of the Method of Characteristics investigated whether there were any advantages to a second order approximation to friction loss, namely basing the friction on the average of the friction terms for points R and P in Equation 3.23 and S and P in 3.25. This introduces an unknown $u_P|u_P|$ term into both characteristics and therefore implies that the simultaneous solution of these equations would require an iterative approach. A numerical sensitivity analysis demonstrated that there was generally no advantage as the rate of change of the flow condition was normally slow enough to support the quasi-steady assumptions introduced above.

However the modelling of unsteady friction has been an important constituent of mainstream pressure surge research over the past 40 years and therefore an enhanced frictional model may be developed based on previous research applicable to pressure surge generated unsteady flow in fluid networks.

The earliest and simplest models to include unsteady flow in a frictional representation were due to Daily et al. (1956) and Carstens and Roller (1959) who developed an additional empirical term to be added to the friction factor,

$$f_{unsteady} = f_{steady} + 0.449 \frac{D}{V^2} \frac{\partial V}{\partial t} \tag{3.39}$$

with frictional loss defined by $\Delta p = \dfrac{4 f \rho L V^2}{2D}$

Abreu and Almeida (2000) commented that this model would be limited due to its neglect of a time evolution of the unsteady flow, however this model does have the advantage of being simple to introduce into an MoC simulation with no major increase in computational time for a complex network.

Vitkovsky et al. (2004) presented a review of the later methods developed to represent unsteady friction in both laminar and turbulent flows. In the case of laminar flows the seminal work of Zilke (1968) is recognised as the basis of all later unsteady frictional research. Zilke defined for the first time the frictional terms necessary to include the time evolution of unsteady flow in the determination of the applicable friction term in a form readily implemented within an MoC

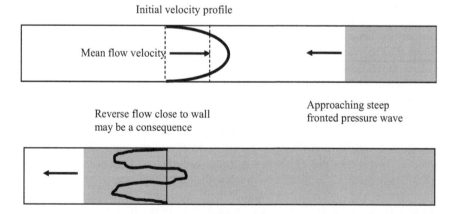

Figure 3.5 The effect of a steep fronted transient on the assumed steady flow velocity profile across a duct

simulation. The Zilke solution considers two-dimensional axi-symmetric laminar flow and thus represents the secondary internal flows that can be induced during the passage of a transient, Figure 3.5. The Zilke solution for unsteady frictional pressure loss is a convolution of a weighting function and cumulative previous fluid accelerations,

$$\Delta p = \frac{4 f \rho L V^2}{2D} + \frac{16 \rho L}{D^2} \int_0^t \frac{\partial V}{\partial t^*} W_0 \left(t - t^* \right) dt^* \tag{3.40}$$

where t^* is non-dimensional time,

$$t^* = \frac{4 \upsilon t}{D^2} \tag{3.41}$$

and W is a weighting factor developed by Zilke.

The Zilke model includes the two-dimensional pressure and flow gradients present within the unsteady flow. The weighting factors, W, were derived by assuming a constant kinematic viscosity across the pipe that remained unaffected by the passage of the transient and may be determined from the following coefficients.

$$W\left(t^*\right) = \sum_{j=1}^{6} m_j \left(t^*\right)^{\frac{j-2}{j}} \quad \text{for } t^* \leq 0.02$$

$$W\left(t^*\right) = \sum_{j=1}^{5} e^{-n_j t^*} \quad \text{for } t^* > 0.02$$

where

$m_j = 0.282095, -1.25, 1.057855, 0.9375, 0.396696, -0.351563, j=1,6$

and

$n_j = 26.3744, 70.8493, 135.0198, 218.9216, 322.5544, j=1,5$

Later work reviewed by Vitkovsky et al. (2004) extended Zilke's work to turbulent flows and introduced various simplifications to reduce the time consuming determination of the frictional terms. Zilke's model demonstrated close agreement with the experimental measurements of pressure surge in laminar oil flows in small diameter pipes.

Vitkovsky et al. (2004) state that the Zilke integral may be approximated by using the rectangular rule and the acceleration term may be approximated using a central finite difference representation

$$\Delta p = \frac{4 f \rho L V^2}{2D} + \frac{16 \rho L}{D^2} \sum_{1,3,5}^{M} \left[V \left(time - j\Delta t + \Delta t \right) - V \left(time - j\Delta t - \Delta t \right) \right] W \left(t^* \left(j\Delta t \right) \right) \tag{3.42}$$

where $M = time/\Delta t - 1$ and 'time' is the current calculation time, referred to as $t + \Delta t$ in Figure 3.4.

This expression has similarities to the earlier Carstens and Roller expression in that it includes the steady state frictional loss along with a term dependent on the flow local accelerations and the time history of the flow – the latter missing from the earlier expression.

Expansion of the summation in Equation 3.42 shows that the series consists of one unknown term based on local velocity at time $t+\Delta t$ and $M/2$ terms based on historic known values of local velocity.

$$\Delta p = \frac{4f\rho L V^2}{2D} + \frac{16\rho L}{D^2} \sum \Big[\big[V(t+\Delta t) - V(t) \big] W(\Delta t) \Big]_{j=1} +$$
$$\Big[\big[V(t-\Delta t) - V(t-3\Delta t) \big] W(3\Delta t) \Big]_{j=3}$$

(3.43)

In Equation 3.43 the summation is only illustrated for $j=1,3$ but it clearly continues, increasing as time increases. In terms of the bisection method the known terms at $j=3$ onwards would be calculated once at each time step prior to the bisection.[1] The summed truncated series would then be added to the friction term in Equations 3.23 and 3.25.

The number of terms included in the summation increases with time. If the simulation time step is 0.001 and if the simulation runs for 60 seconds then the maximum number of terms included in the summation will increase to $0.5*60*10^3$.

While the quasi-steady frictional approach is sufficient for the majority of building drainage and vent system simulation, there are conditions where an unsteady friction model would be advantageous. A later treatment will illustrate one such case – the modelling of a 10 Hz transient introduced into a network in order to identify changes in boundary conditions following trap seal depletion, Kelly et al. (2008b).

3.5   Boundary condition modelling

Figure 3.6 makes it clear that while velocity, wave speed and pressure at all the internal nodes, $i = 2$ to N, may be determined from the simultaneous solution of the characteristic Equations 3.23 and 3.25, the conditions at entry to, and exit from, a pipelength are only represented by one characteristic at each location, a C^- characteristic at entry at node $i = 1$ and a C^+ characteristic at exit at node $i = N + 1$. The art involved in the application of the Method of Characteristics to simulate transient response is in the determination and expression of suitable boundary equations to be solved with the appropriate characteristic to yield values of u, c and p at each boundary. Without these boundary equations it is clear that the solution cannot proceed beyond $N/2$ time steps, where N is the number of internodal reaches in any pipelength. Note that x is taken as positive from entry to exit; however, in the modelling of vertical stacks, x is taken positive upwards from the sewer connection at the stack base hence the entrained airflow is normally negative.

Thus a boundary condition equation must be provided that is solvable with the one characteristic available at a boundary to yield values of u, c and p and allow

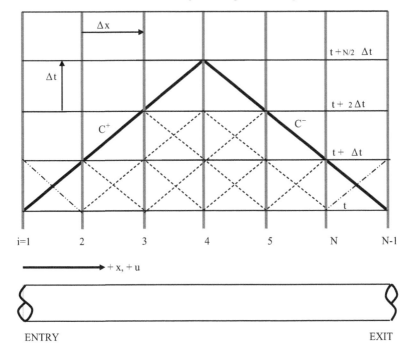

Figure 3.6 Zone of influence of the characteristic equations demonstrating the necessity of boundary equations to allow the solution to proceed

the simulation to proceed as these values are required at each node in the network at all time steps. Generally boundary condition equations may be subdivided into two clearly defined subgroups, namely passive and active boundary conditions.

Passive boundary conditions remain unchanged regardless of the local pressure response of the network. The simplest examples are open or dead ended terminations or junctions of two or more pipes. In these cases it is sufficient in the case of an open end termination to equate local wave speed to that at the prevailing ambient pressure, or for a dead end to equate the local flow velocity to zero, in either case allowing direct solution of the boundary characteristic and the determination of pressure from Equation 3.28.

Active boundary conditions react to the local pressure regime, for example an air admittance valve may be characterised as a dead end so long as the local pressure is positive, however as the local pressure falls below atmosphere the AAV starts to open at some predetermined negative pressure. This stage is followed by an opening phase where the valve loss coefficient reduces as the local system pressure falls further. At some negative pressure level the AAV may be considered fully open and the loss coefficient thereafter remains constant until the local pressure starts to rise and the process is reversed.

Junctions in the stack, vent or system branches may be regarded as either active or passive boundary conditions. In the case of a multi-pipe junction in the dry vent system, the boundary may be seen as passive as the required solution is based wholly on the system geometry that defines how many pipes are considered to terminate at the junction, and hence are represented by C^+ characteristics, and how many are considered to start at the junction, represented by C^- characteristics, together with the equations of flow continuity and pressure equivalence at the junction.

However junctions in the wet stack, where one of the joining pipes is a discharging branch, may be seen as an active boundary as the severity of the pressure loss encountered by the entrained air drawn through the water film established across the stack from the discharging branch is dependent on the discharge flow and therefore on time.

Any system simulation has to incorporate this range of boundary conditions and incorporate a mechanism to determine, by monitoring the system response and the applied flowrates to the network, the appropriate boundary condition that must be applied at any location.

Similarly, as will be shown in the later treatment of transient control and suppression, it is necessary to include boundary conditions that may not be activated if the system pressure regime does not cross preset boundaries. For example the active boundary representing an air admittance valve may not be utilised unless the system pressure falls below a certain threshold and similarly the boundary conditions representing a variable volume containment device (PTA) may not be activated unless the system pressure exceeds a preset positive pressure at the device location.

Boundary condition choice and definition is therefore at the centre of any Method of Characteristics application.

The process of building an active boundary condition equation may be demonstrated by reference to the response of an appliance trap seal to the system side incoming air pressure transients arriving at the water–air interface. This boundary condition must be capable of representing the displacement of the trap seal, including possible loss of seal water as a result of either positive or negative incoming transients and the partial loss of seal if the water column is sufficiently displaced to allow an air path to or from the appliance – an effect sometimes known as 'bubble through', which may not be discernable once the transient event is completed. Similarly the boundary condition should cater for changes to the room ambient pressure applied on the trap seal appliance side.

In the case of a branch terminated by an appliance trap the appropriate boundary condition is provided by the application of Newton's second law to the trap water seal. This expression is solved with the available C^+ characteristic. Note this characteristic is appropriate, as the normal convention is to measure distance positive from the base of the system stack so that the trap becomes an exit boundary condition.

Figure 3.7 illustrates the trap water seal at two instants in time separated by the simulation time step, governed by the Courant condition and defined as

$$\Delta t = \frac{\Delta x}{\left(u + c\right)_{MAX}}$$

The forces acting on the displaced water column may be defined as follows.

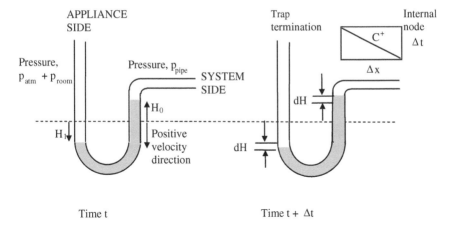

Note trap diameter, D and wetted perimeter P. The trap seal water length L may vary dependent upon the applied pressure transient regime as the trap seal is lost.

Figure 3.7 Definition of the trap seal termination illustrating trap deflection across a time step

1   The differential between room air pressure and the transient pressure in the branch at the trap interface,

$$\text{Term T1} = A\left[0.5\left(p_{pipe,t+\Delta t} + p_{pipe,t}\right) - \left(p_{atm} + p_{room}\right)\right] \tag{3.44}$$

2   The differential in water surface levels at time $t+Dt$ may be expressed as

$$\text{Term T2} = \rho g A\left(H_0 - H_1 + 2dH\right) \tag{3.45}$$

3   The frictional resistance to trap seal oscillation may be expressed as

$$\text{Term T3} = \tau L P = 0.5\rho f L P V_{trap}^2 \tag{3.46}$$

where the shear stress t is represented via an empirical friction factor.

4   The mass acceleration term may be expressed as

$$\text{Term T4} = \rho L A \frac{V_{trap,t+\Delta t} - V_{trap,t}}{\Delta t} \tag{3.47}$$

to yield a boundary condition expressed as

$$F = T1 + T2 - T3 - T4 = 0 \tag{3.48}$$

to be solved with the available C^+ characteristic expressed as

$$u_{trap,t+\Delta t} = K1 - K2c_{trap,t+\Delta t} \tag{3.49}$$

where the constants K1 and K2 are dependent on flow and frictional term values known at time t and, to allow for trap diameters smaller than the connected branch,

$$u_{trap,t+\Delta t} = \left(\frac{D_{trap}}{D_{branch}}\right)^2 V_{trap,t+\Delta t} \tag{3.50}$$

where V is the trap water velocity. Solution may be achieved via the bisection method applied to the function $F = T1 + T2 - T3 - T4$ between suitable air pressure limits. The C^+ characteristic is included via Term 4 once the wave speed has been determined from the assumed trial air pressure, p_X as

$$c_{trap,t+\Delta t} = \sqrt{\frac{\gamma p_X}{\rho_{atm}\left(\dfrac{p_X}{p_{atm}}\right)^{\frac{1}{\gamma}}}} \tag{3.51}$$

Trap seal oscillation may be identified as laminar flow based on the fluid velocity and Reynolds number, Re, of the imposed flow. The friction factor in Term 3 may therefore be represented by the steady flow laminar friction factor relationship

$$f = \frac{16}{Re} = \frac{16\upsilon}{V_{trap}D_{trap}} \tag{3.52}$$

However the flow conditions in the trap are not steady as the trap seal will oscillate in response to changes in system air pressure and a damped oscillation will occur following cessation of the system transient pressure propagation. An improved frictional representation is therefore required. A study of trap seal response would suggest that the steady state friction assumption would severely underestimate the frictional forces acting on the oscillating column due to the possibly rapid reversal of flow direction within the trap inherent in its oscillation. There is a requirement to include the local flow acceleration in the derivation of an appropriate friction term. The treatment of unsteady friction presented above can be applied directly to the trap seal water motion, utilising Zilke's expressions for laminar unsteady flow. As already stressed the advantage will be dependent upon the rate of change of the applied transient at the water to air system side interface and it will be shown later that for the typical transient rise times derived from appliance discharge a quasi-steady model is sufficient.

The solution of the C^+ characteristic with the boundary condition based on the trap seal equation of motion delivers in the first instance the pressure, air velocity and wave speed at the branch termination as well as the water levels in the trap. It will be seen that trap seal water may be lost to the branch if a negative incoming transient displaces the water column towards the system side, thus the mass of the trap seal water is time dependent and its variation must be included in the model.

Similarly a positive transient may 'drive' the water column upwards into the appliance and reveal an air path through the trap. This may be included in the boundary condition by monitoring the water levels in each 'arm' of the trap and setting the system side pressure to room pressure once the airpath is revealed.

This example illustrates that a boundary condition may in fact be represented by a family of boundary equations, the selection depending upon the local conditions – in this case the displacement of the trap seal water column. Other common boundary conditions are illustrated in Figure 3.8 and also presented in Table 3.2, as either passive or active boundary conditions.

3.6   Transient driving functions

Generally, in the modelling of pressure transient phenomena, it is accepted that the driving functions are represented by boundary conditions that are time

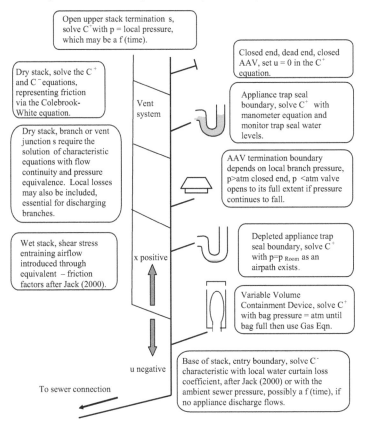

Figure 3.8 Boundary condition equations solved with the appropriate characteristic for a range of boundaries, both internal and terminal, found in a building

Table 3.2 Active and Passive boundary conditions within a building drainage and vent system

Boundary	Active or Passive Boundary	Boundary Condition	Available Characteristic Equation	Boundary Equations
Open end exit	Passive	Open to atmosphere or a known ambient pressure	Exit C^+ solved with local wave speed based on p = ambient pressure	C^+ Equation 3.23 Wave speed Equation 3.27 Pressure Equation 3.28
Concentrated separation loss at open exit	Passive	Open to atmosphere or a known ambient pressure but with a restricted exit generating a concentrated local pressure loss	Exit C^+ solved with local wave speed based on $p = p_{ambient} + \dfrac{1}{2}\rho u\lvert u\rvert$ note abs. value ensures pressure loss opposes flow	C^+ Equation 3.23 Wave speed Equation 3.28 and $p = p_{ambient} + \dfrac{1}{2}\rho u\lvert u\rvert$
Dead end exit	Passive	Exit C^+ solved with $u_{local} = 0.0$	Exit C^+ solved with $u_{local} = 0.0$	Solve with available C^+ Equation 3.23
Stack, Vent or branch junction with no local separation losses	Passive	Consider n pipes starting and m pipes ending at a junction. Unknowns are p, u and c for each entry node 1 and each exit node N+1, a total of $3(m+n)$ unknowns requiring $3(m+n)$ solvable equations	m C^+ characteristic at each of m pipe terminations and n C^- characteristic at each of n pipe entries. plus a continuity of flow equation $$\sum_i^{m+n} uA = 0$$ and $m+n-1$ pressure equivalence equations with no loss coefficient of the form $$p_1 = p_2 = p_3 = \ldots = p_{m+n},$$ $m+n$ pressure wave speed equation provide the remaining equations	Solve the available C^+ and C^- characteristics simultaneously with the continuity equation and the $m+n-1$ pressure equivalence expressions to yield c and u and then p from Equation 3.28
Stack, Vent or branch junction with local separation losses based on the airflow into each branch	Passive	Similar to the 'no local loss' junction except that the pressure equivalence expressions now have to include an empirical loss coefficient for each route	The pressure equivalence equations with local loss coefficients take the form $$p_1 = p_2 + k_{1-2}u_1^2 = p_{m+n} + k_{1-m+n}u_1^2$$	Solve the available C^+ and C^- characteristics simultaneously with the continuity equation and the $m+n-1$ pressure equivalence expressions including the loss coefficients

Boundary	Active or Passive Boundary	Boundary Condition	Available Characteristic Equation	Boundary Equations
Wet stack junction with a discharging branch, limited to the junction of three pipes	Active	The discharging branch results in a water curtain being formed across the vertical stack that introduces a local pressure loss as the entrained airflow is drawn through. This loss is empirical and values based on Jack (2000) and O'Sullivan (1974) are employed	The pressure loss encountered by the entrained airflow in passing from an upper stack section, pipe 3, to a lower stack section, pipe 1, may be written as $p_1 - p_3 = \frac{1}{2}\rho K_{/(Qw, stackgeometry)}u_1^2$ As wet stack junctions limited to three pipes due to the available empirical data, 9 equations are required, 3 characteristics, flow continuity and 2 pressure equivalence expressions and 3 pressure wave speed expressions	At a wet stack junction the lower stack section is deemed to end and the branch and upper stack section start. This yields 1 C^+ and 2 C^- equations. The branch and upper stack pressures, p_2 and p_3 'are assumed' equal as both are above the water curtain while the pressure in the lower stack section, p_1 is given by $$p_1 = p_2 - \frac{1}{2}\rho K_{junction}u_1^2$$
Open end exit with variable external applied pressure, inc. wind shear	Active	Exit C^+ solved with local wave speed based on $p = f(time)$	Exit C^+ solved with local wave speed based on $p = f(time)$	C^+ Equation 3.23Wave speed Equation 3.28 using time dependent p
Stack base	Active	Water curtain forms across the entry into the horizontal drain and presents a local loss applied to the entrained airflow	Entry C^- solved with local empirical loss coefficient expression that depends upon stack geometry (Passive) and also the applied water downflow (hence seen as an Active boundary)	Solve C^- Equation 3.25 with the stack base water film loss coefficient equation taken as $$p_{local} = K_{Qw}\frac{1}{2}\rho u_{local}|u_{local}|$$ as K increases with the water flow Q_w
Stack base Surcharged	Active	Excessive water flow effectively blocks the airpath from the stack to the horizontal drain	Airflow is reduced to zero, C^- equation solved as for a dead end by setting local velocity to zero. Water downflow monitored to indicate re-instatement of an airpath to the horizontal drain	Entry C^- Equation 3.25 solved with the local velocity $u_{local} = 0.0$
Vertical stack offsets	Active	Excessive water flow effectively blocks the airpath through offsets that are often necessary in vertical stacks due to building design	Airflow is reduced to zero, C^+ and C^- Equation solved downstream and upstream of the offset as for a dead end by setting local velocity to zero. Water downflow monitored to indicate re-instatement of an airpath to the horizontal drain.	Exit C^+ Equation 3.23 and Entry C^- Equation 3.25 solved with the local velocity u_{local} set to zero 0.0. Note downstream pressure surge −ve while upstream surge +ve

Continued …

Table 3.2 continued ...

Boundary	Active or Passive Boundary	Boundary Condition	Available Characteristic Equation	Boundary Equations
Stack base subject to sewer generated pressures, possibly remote from system with no water downflow	Active	Analogous to the wind shear boundary applicable to an open topped vent stack	Entry C^- solved with local wave speed based on $p = f(\text{time})$. This assumes that the sewer pressure variation is known	C^- Equation 3.25 Wave speed Equation 3.28 using time dependent sewer pressure
Stack base subject to sewer generated pressures while carrying water downflow from discharging appliances	Active	The pressure differential across the water curtain formed across the entry into the horizontal drain includes the effect of a varying downstream sewer side pressure	Entry C^- solved with local empirical loss coefficient expression that depends upon stack geometry and downflow and an additional term to represent the time dependent downstream reference pressure	Solve C^- Equation 3.25 with the stack base water film loss coefficient equation taken as $$p_{local} = p_{SEWER} + K_{Qw}\frac{1}{2}\rho u_{local}\left|u_{local}\right|$$ as K increases with the water flow Qw. Monitor water downflow to identify potential surcharge and change boundary condition appropriately
Air Admittance Valve (AAV). The valve entry loss coefficient depends upon the degree of valve opening which in turn depends on the pressure differential across the AAV. Iterative solution of the valve loss equation with the C^+ equation is necessary until valve fully open	Active	Dependent on local pressure, cf. opening pressure and fully open pressure levels 1. $p_{local} > p_{open}$ 2. $p_{local} < p_{open}$ 3. $p_{local} < p_{fully\ open}$	Solve available C^+ characteristic with the valve entry loss coefficient – this depends on the local pressure differential across the AAV, one of three conditions may apply as shown	Solve valve loss equations below with Equation 3.23 1. Treat as a dead end exit, set $u_{local} = 0$ 2. Solve C^+ with the AAV loss coefficient equation taken as $p_{local} - p_0 = K_p\frac{1}{2}\rho u_{local}\left|u_{local}\right|$ note K reduces as valve opens 3. Entry loss coefficient becomes constant once valve fully open

Boundary	Active or Passive Boundary	Boundary Condition	Available Characteristic Equation	Boundary Equations
Appliance water seal trap	Active	1. No trap seal loss 2. Trap seal depleted to allow an airpath from the system into habitable space 3. Trap seal lost completely	Exit C^+ solved with the equation of motion for the trap water column. Monitor water column length to determine if seal depleted	1. C^+ Equation 3.23 Trap equation of motion 3.48 2. Set p_{local} = atmospheric pressure and solve with C^+ Equation 3.23 3. Set p_{local} = atmospheric pressure and solve with C^+ Equation 3.23
Variable Volume Containment Device. A variable volume containment device that expands to its full volume to provide an alternative airpath and attenuate transient	Active	Dependent on local pressure the containment volume will open to the drain connection and then expand until full when pressurisation occurs and the device's efficiency is reduced. 1. p_{local} < atmosphere 2. p_{local} > atmosphere but bag volume < fully expanded 3. p_{local} > atmosphere but bag fully expanded	Exit C^+ Equation may be solved with the appropriate bag pressure dependent entry condition	1. C^+ Equation 3.23 with the local velocity $u_{local} = 0$ 2. Set p_{local} = atmospheric pressure and solve with C^+ Equation 3.23. Monitor inflow rate to determine if bag fully expanded at atmospheric pressure 3. Solve C^+ Equation 3.23 with the gas law applied to the expanded bag, $p\mathrm{Vol}$ = constant

dependent. Obvious examples drawn from the wider pressure surge application of the Method of Characteristics include valve closure, resulting in upstream positive surges and possible downstream cavitation, column separation and re-surge, or pump operation or emergency shut down, planned or as a failure condition, that may lead to column separation and damaging re-surge pressures.

In the application of the Method of Characteristics to the low amplitude air pressure transients found in building drainage and vent systems, this is also true to some extent, the surcharge experienced at the base of a stack or at a stack offset may well bring the entrained airflow to rest and generate the traditional Joukowsky pressure transient. Similarly the sudden closure of an AAV when the pressure rises above atmosphere will generate a negative transient as the inflow through the valve is rapidly cut off.

However, the main cause of low amplitude air pressure transient propagation in building drainage and vent systems is the shear force interaction between the annular downflow of appliance discharges and the air core within the system's vertical stack. This effect generates the entrained airflow within the system and any changes in the water downflow, whether increases due to further appliance discharges, or decreases as the system becomes quiescent, propagate transients throughout the network.

Thus in the case of building drainage and vent systems the most important boundary condition is the shear force relationship between the annular falling water flow and the air within the vertical stack. The development of a suitable model was based on extensive site testing and the resulting expressions linking traction to water flow and system geometry are wholly empirical.

3.7   Traction forces acting on the air core within a building drainage and vent system

The characteristic Equations 3.23 and 3.25, applicable across two internodal reaches in the vertical stack, include frictional terms, f_R and f_S, normally expected to represent the frictional resistance to fluid flow and provide a degree of transient attenuation.

$$u_P - u_R + \frac{2}{\gamma - 1}(c_P - c_R) + 4 f_R u_R \mid u_R \mid \frac{\Delta t}{2D} = 0$$

$$u_P - u_S - \frac{2}{\gamma - 1}(c_P - c_S) + 4 f_S u_S \mid u_S \mid \frac{\Delta t}{2D} = 0$$

However, in the case of an entrained airflow within a vertical stack carrying an annular water downflow, it can be seen that these terms may be modified to represent the shear forces acting on the air core to entrain a central airflow. The development of such an empirical approach to the traction mechanism within building drainage and vent systems was the objective of a major site test and analysis programme, Jack (1997, 2000), Swaffield and Jack (1998). Figure 3.9

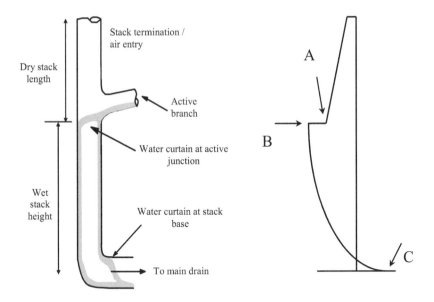

Figure 3.9 Single stack pressure profile (single discharge)

illustrates the long-established understanding of the pressure regime within a single stack drainage system.

The key pressures defining the mechanism may be identified as

> Pressure 'A': The suction pressure at the base of the dry stack due to the dry stack friction and any entry loss.
> Pressure 'B': The local pressure loss at a discharging branch junction, dependent on the entrained airflow rate and the shape and coverage of the water film formed at the junction.
> Pressure 'C': A (positive) back pressure frequently generated at the base of the vertical stack as the entrained airflow passes through the water curtain formed as the annular downflow enters the horizontal drain.

Jack (2000) determined these pressure levels for a range of stack diameters, roughness values, discharge junction heights and applied flowrates to allow a determination of the shear forces generating the entrained airflow by employing Buckingham's π theorem (1914).

By summing the 'absolute' magnitude of each of these constituent pressures, the 'traction work done' by the falling annular water film (whether generated by single or multiple discharge flows) was calculated by application of the steady flow energy equation that provides an audit across the control volume encompassing the drainage network, a concept considered by Lillywhite and Wise (1969). As pressure losses due to friction may be represented by Darcy's equation, where

$$\Delta P_f = \frac{4\rho f L V^2}{2D} \tag{3.53}$$

substituting stack parameters allows of the shear forces to be defined by a pseudo-friction factor, f, over the wet stack height H_w

$$f = \frac{2D_a |\Delta P|_{sum}}{4\rho_a H_w (V_a - V_{t1}) abs(V_a - V_{t1})} \tag{3.54}$$

where $(V_a - V_{t1})$ represents the velocity difference between the water surface and the mean airflow velocity, the usual friction factor representation. This is the equivalent of bringing the annular film to rest by the superposition of an equal and opposite velocity and allows a frictional representation to be included in the characteristic equations, resulting in a 'negative friction factor' or pseudo-friction factor and a pressure rise in the airflow direction. Crucially this model allows predictions that need not be limited to single point discharge conditions as through the use of a variable friction factor term allows combined discharge flows and their effect on the entrainment of air to be modelled.

Figure 3.10 summarises this approach and confirms the model's ability to simulate either single or combined discharges and demonstrates that the friction factor sign notation conforms to the rise or fall in air pressure in the airflow direction experienced within various sections of the stack.

Within the dry stack, the entrained airflow velocity obviously exceeds that of the stationary stack wall and so the friction acting between the entrained airflow and the pipe has a positive sign resulting in a drop in air pressure.

Within the wet stack, downstream of the upper 'active' branch connection, the interface velocity of the water is, under normal circumstances, greater than the mean velocity of the entrained air. This results in a 'negative' friction factor term which gives a corresponding rise in air pressure.

However, when the stack is subject to multiple flows which combine or 'mix' within the vertical stack, it is possible that the increased discharge flow rate present in the lower section of the stack can induce or entrain a significantly higher airflow such that the resultant air velocity in the upper section of the wet stack is greater in magnitude than the surface velocity of the annular water film. Effectively the entrained airflow is drawn past the falling annular water film. This then gives a positive friction factor and a corresponding pressure drop in the airflow direction. The resulting pressure drop in this section of the wet stack is accommodated in the model by introducing an increased local relative roughness.

The water film velocity to be used in the frictional representation of the water to air shear force is the water surface velocity, not the terminal velocity of the annular flow. Figure 3.11 illustrates the likely water and airflow velocity profiles.

The assumed flow regime shown in Figure 3.11 demonstrates a linear water velocity profile; however, it is likely that this is not the case and that the surface velocity may be as high as three times that of the mean terminal water velocity and that the profile shape across the annular film would be a parabolic, rather than linear, function. Thus the frictional terms included in Equations 3.23 and 3.25 become

$$u_P - u_R + \frac{2}{\gamma - 1}(c_P - c_R) + 4 f_R (u_R - n V_{Term} \mid u_R - n V_{Term}) \mid \frac{\Delta t}{2D} = 0 \qquad (3.55)$$

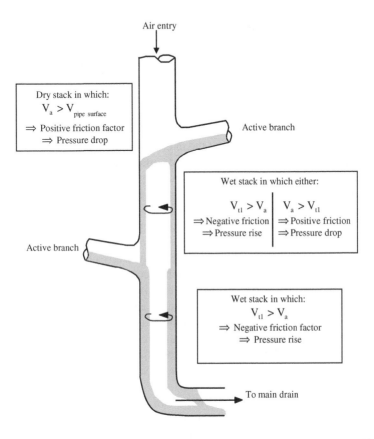

Figure 3.10 Definition of friction factor application through single stack system subject to multiple simultaneous discharges

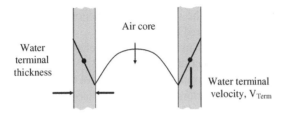

Figure 3.11 Water and air velocity profiles

$$u_P - u_S - \frac{2}{\gamma - 1}(c_P - c_S) + 4f_S(u_S - nV_{Term}) \mid u_S - nV_{Term} \mid \frac{\Delta t}{2D} = 0 \qquad (3.56)$$

where n is a suitable multiplier to account for the annular water film velocity profile and V_{Term} is the annular film terminal velocity, dependent upon water flow rate, and stack diameter and roughness.

The identification of the 'pseudo-friction factor', derived from an application of the steady flow energy equation to the site data, Jack (1997), has allowed the development of the driving functions that enable AIRNET to simulate multiple branch discharges to the vertical stack and, as will be shown later, multi-stack networks. However, while these results and the driving equations in AIRNET are based on extensive site tests and laboratory work, these tests, in common with all the historic published stack research, utilised clean water as the annular flow medium. In many drainage network applications the annular water film may contain surfactants, derived from washing machine and dishwasher discharge, that may affect the driving values of the interface shear stress.

MacLeod (2000) considered the effect of surfactant addition on both entrained airflow and stack pressure profile. While detergents in general are classified as surfactants – i.e. they will effect the shear forces between the air core and the annular water film – MacLeod considered two specific types, namely hydrophilic (water compatible) and hydrophobic (water repellent). The water compatible group includes anionic (a negatively charged group generally found in domestic soap powders, tablets and liquids), cationic (a positively charged group found in fabric conditioners) and non-ionic (an uncharged group that is the second most widely used in washing powders). The presence of a surfactant modifies the water surface tension and the shear force applied to the air core. Figure 3.12 illustrates an increase in air entrainment with both anionic and non-ionic surfactants, implying a reduced surface tension, an increased shear force, a higher annular film terminal velocity and reduced film thickness. Addition of a cationic surfactant reduced the entrained airflow, implying a reduced shear force, terminal annular flow velocity and increased film thickness.

Figure 3.13 presents compatible minimum air pressure values within a 100 mm glass vertical stack at a constant water flow of 2 litres/second with a range of stack entry loss coefficients generated by progressively reducing the upper vent diameter. As the stack blockage approaches 100 per cent so the minimum pressure falls further, the relative minimum pressure values confirming the air entrainment effect of each surfactant type in Figure 3.12. MacLeod found that foaming in the lower stack occurred as the stack entry restriction approached 100 per cent. (Foaming is extensively covered by ASPE 2002.)

The effects of surfactants may therefore be characterised as a dependency within the shear driving function in Equations 3.55 and 3.56. The value of n, the annular terminal velocity multiplier, would have to vary with surfactant type and concentration up to the saturation level identified by Campbell and MacLeod (1999). Unfortunately there is insufficient data to generalise this effect across stack diameter and roughness.

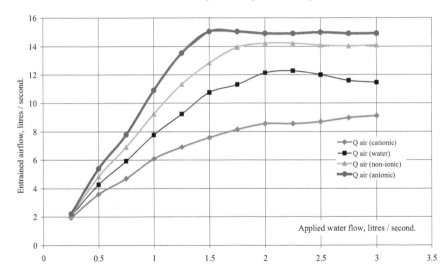

Figure 3.12 Air entrainment as affected by surfactant action in a 100 mm diameter glass vertical stack subjected to an increasing applied water downflow

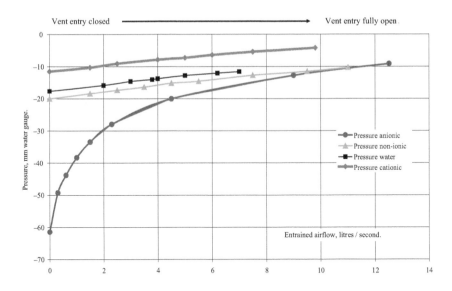

Figure 3.13 Suction pressures in a vertical stack, subject to variable entry loss, as modified by the presence of surfactants

The mechanism of air entrainment is generally agreed to be dependent upon the shear force between the annular water film and the air core. Wyly and Eaton (1961) and Pink (1973b) concluded that the majority of the water downflow fell as an annular flow, Pink in particular finding little evidence of droplet water flow in the air core. More recently it has been postulated that water droplet flow does occur and contributes to the overall air entrainment; Campbell and MacLeod (1999) and Cheng et al. (2009) have both contributed to this discussion. It is likely that droplet flow does occur, however it would be difficult to determine the relative importance of the two shear processes. Campbell and MacLeod (1999) postulate an annular shear force dependent upon a V^2 relationship between the air core and the annular surface velocity and a droplet drag force dependent upon a V^2 relationship between the droplet velocity and the surrounding airflow velocity. The approach followed by Jack (2000) effectively captures both these forces as an application of the steady flow energy equation would combine both within the 'pseudo-friction' factor identified earlier.

Equations 3.55 and 3.56 therefore provide the essential driving function that allows the MoC simulation to generate the pressure recovery that characterises the airflow in the wet stack and allows the simulation to encompass both multiple branch discharges to the stack and variable entry losses.

3.8   Concluding remarks

This chapter has introduced the Method of Characteristics as a numerical finite difference scheme ideally suited for use in the solution of the St Venant equations and the modelling of pressure transient propagation within fluid networks as a consequence of system operating point change. Linked to the application of the Courant criterion, which determines the simulation time step, the MoC solution has become, since the 1960s, the industry standard technique to solve and simulate transient events. This is as true of the treatment of low amplitude air pressure transients in building drainage and vent systems as it is in all other fluid transient applications.

This chapter has also introduced the fundamental concept of the boundary equation. As only one characteristic can exist at a pipe entry or exit it becomes necessary to provide a suitable set of boundary equations to cover all the possible permutations of boundary conditions. The simulations to be developed and demonstrated in later chapters will operate on the basis that the simulation itself identifies the appropriate boundary condition at any exit or entry based on the predicted local conditions, examples already given include the air admittance valve and the PTA pressure transient control devices.

The introduction of a finite difference scheme to solve the St Venant equations also allows the introduction of frictional resistance into the flow equations. Generally a quasi-steady approach is deemed sufficient as the rise time of most transients will exceed the pipe periods appropriate to the systems simulated. However, the Method of Characteristics solution is sufficiently robust to allow the inclusion, when necessary, of unsteady frictional modelling based on specialist studies over the past 40 years. An application in the modelling of system response to sinusoidal excitation in a practical system designed to identify in a non-invasive manner dry trap seals will be demonstrated in a later chapter.

In addition the role of the shear forces present between the annular water film in a vertical stack and the entrained airflow in the core has been discussed and an approach that introduces a pseudo-friction factor has been demonstrated that allows a full modelling of the pressure regime within a drainage network.

Thus this chapter has provided the mechanisms that will allow the difficulties identified in Chapter 2, relating to the complexity of the pressure transient response and the requirement to include frictional resistance, to be overcome. In addition the Method of Characteristics solution described will allow a wide range of low amplitude air pressure transient conditions to be simulated in order to allow a better understanding of system operation and will therefore provide a tool in the development of both improved design codes and strategies to control and suppress low amplitude air pressure transients in building drainage and vent systems.

Chapter 4 will develop the solutions introduced by demonstrating their application to a wide range of common transient events within typical drainage and vent systems.

Note

1 **Bisection method**
   An iterative technique to solve for boundary conditions.

   Set upper and lower limits to the solution, xu and xl, Set acceptable level of $f(0)$ to approach zero.

   A   Set trial value as $xt=0.5(xu+xl)$,
        Calculate function value $f(xt)$; IF $f(xt)>0$, set $xu=xt$; IF $f(xt)<0$, set $xl=xt$
        Return to A and recalculate xt and repeat until $f(xt)<f(0)$
        Solution $x=xt$

4 Simulation of the basic mechanisms of low amplitude air pressure transient propagation – AIRNET applications

Previous chapters have presented the fluid mechanisms that lead to the generally accepted air pressure regimes encountered within building drainage and vent systems, regimes that are predicated on the entrainment of an inflow by the applied appliance discharges and the subsequent interaction of that entrained airflow with the component parts of the drainage network and other system flows. In addition, previous chapters have presented the development of pressure transient theory and practice over the 100 years since the seminal work of Joukowsky, a multi-disciplinary endeavour with a full range of international contributors, as well as the development of the Method of Characteristics as the industry standard for surge analysis following the contributions of Streeter and Fox in the 1960s.

The application of the Method of Characteristics to the solution of the St Venant equations of continuity and momentum within the context of the current text has also been presented with the emphasis that low amplitude air pressure transient propagation in building drainage and vent systems conform in all respects to the fundamental principles of pressure surge analysis. The emphasis has been placed on both the development of simulation techniques that allow the solution of the governing equations as well as the development of unique boundary conditions applicable only within building drainage networks. In particular the identification of driving functions to allow the shear interaction between the falling annular water film in vertical stacks and the entrained air core has been central to the simulations to be presented here and in later chapters. Based on extensive fieldwork these relationships allow the MoC solutions to apply and provide a basis for the air pressure regime predictions to be discussed in this chapter. Individual boundary conditions have already been developed for the wide range of terminal conditions found in a typical network, including 'passive' boundaries such as dead ends, open stack terminations and pipe junctions and active boundary conditions such as water trap seal column oscillation, operation of air admittance valves (AAVs) and flexible variable volume containment devices, currently only represented by the P.A.P.A.™ device invented by Swaffield and Campbell and marketed by Studor Ltd.

This chapter will present a series of simulations that allow the fundamental transient processes to be identified and the influence of system parameters and control and suppression devices to be modelled. The text will present AIRNET simulation results in graphical form to allow comparison between drainage network designs and component choices.

4.1 Drainage and vent system design and the simulation of the pressure regime in each common system type in response to multiple appliance discharge

Historically there have been four main system design options. From the 1930s the one-pipe system was introduced to facilitate the internal mounting of drainage networks in tall buildings. Figure 4.1(a) illustrates such a system where each appliance is connected via a vent to a dry vent stack that extends the full height of the building, as well as to a wet stack that connects each appliance eventually to the sewer. The wet stack also terminates at roof level in an open termination. In order to reduce system pipework demands in the UK the single stack system was introduced from the 1960s; here, as shown in Figure 4.1(b), there is no parallel dry vent stack. The single stack system was challenged for applications above 30 storeys and the modified one-pipe system was introduced with the wet stack cross-connected to a parallel dry vent stack with no vent connection to each appliance, Figure 4.1(c). Current design guidance suggests that the vent stack should be of smaller diameter than the wet stack, e.g. 100 and 150 mm diameter wet stacks being paired with 50 and 80 mm diameter vents.

From the 1980s onwards the use of air admittance valves has become more accepted, as shown in Figure 4.1(d) and more recently the use of air admittance valves in combination with flexible variable volume containment devices (VVCD), that provide positive air pressure attenuation, has been introduced as a form of Active Control to replace or supplement the Passive Control approach represented by the vent stack approach, Figure 4.1(e). These schematic representations of the drainage networks mentioned will be used throughout this chapter to demonstrate the application of the AIRNET simulation. Each vertical pipe section is 4 m in length and the branches leading to traps, dead ends, AAVs and VVCDs are all of 2 m length. The vent stack pipes are again of 4 or 8 m section length. Pipe diameters will be discussed in each application. Appliance discharges to the stack will be assumed for those branches leading to trap seals. Pipe numbers are shown in Figures 4.1(a) to (e) where they change or where additional pipe sections are added, thus the AAV and VVCD equipped single stack system has the same pipe numbering as the initial single stack. The one-pipe and modified one-pipe systems have additional pipe sections and these are identified. It should be noted that these demonstration networks have been designed specifically to allow comparison of the effect of various parameters, such as individual appliance venting compared with AAV introduction, and are therefore not meant to be taken as examples of installed systems.

The one-pipe, single stack and modified one-pipe will be compared by simulating the effect on trap seal retention of a multiple inflow condition where flow is introduced through the lower four trapped branches. The earlier descriptions of the air pressure regime in the stack suggests that, following a

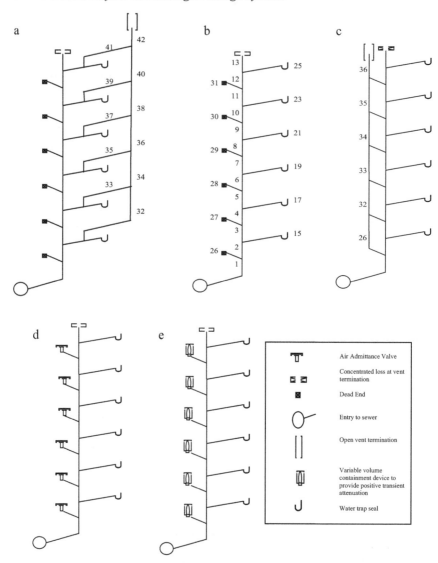

Figure 4.1a) One pipe network (1930); *b)* Single stack (1960); *c)* Single stack modified one pipe (1970); *d)* Single stack with distributed AAVs (pipe numbers as single stack); *e)* single stack with variable volume containment devices (pipe numbers as single stack)

fall in air pressure down the dry portion of the stack and a concentrated pressure drop due to minor losses incurred as the entrained airflow is drawn through the water sheet formed across the stack at a discharging branch, air pressure is regained due to the shear forces acting between the annular water film and the air core. The water curtain formed as the annular water film translates into free surface flow at the stack base provides a final reduction in pressure to that prevailing in the sewer.

A successful drainage design will minimise the trap seal oscillations and any trap seal depletion as a result of the wholly expected air pressure transient propagation that will accompany changes in the applied water discharge to the network. On that basis the first comparison between the systems illustrated will be based on trap seal retention as a result of increasing the four-point inflow to the stack.

The AIRNET simulation will also be used to compare the pressure regime down the stack for each system design at a particular instant in time, the case chosen being representative of conditions when the applied water discharge had increased to 1 litre/second from each of the four branches.

Figure 4.2 illustrates the expected air pressure regime within the single stack system. Pressure falls down the dry stack to the first discharging junction where there is evidence of a concentrated loss. Below this and the lower discharging branches the stack pressure reduces at a much decreased rate due to the mitigating effect of the water to air core shear. Below the lower discharging branch the air pressure recovers and a back pressure above atmosphere is predicted at the stack base before the annular flow becomes free surface flow in the sewer connection. This is entirely as expected and confirms the ability of the simulation to model the experimental results obtained over many decades. The ability of the simulation to deliver this agreement in a multiple inflow case demonstrates its versatility.

Figure 4.3 demonstrates the same general form as the single stack result with two important differences. The first obvious difference is that the air pressure in the stack is reduced from 40 mm water gauge below atmosphere to 10 mm water gauge, thus providing a more benevolent regime for the appliance trap seals.

Pressure profile over the single stack height with multiple appliance discharges.

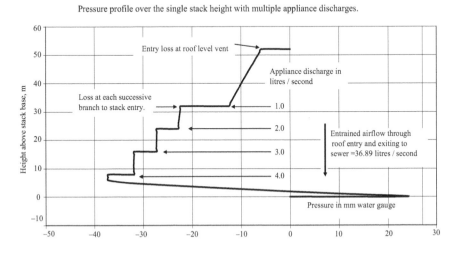

Figure 4.2 Expected air pressure regime within the vertical stack of a single stack system subject to four individual discharging branches

This would suggest that the modified one-pipe system is superior to the single stack system, however the extra piping material required is clear and explains the attraction of the simpler single stack design.

Figure 4.4 demonstrates that the added complication of providing a separate vent connection from each appliance to the vent stack provides little advantage when compared with the simplified modified one-pipe network. However it would

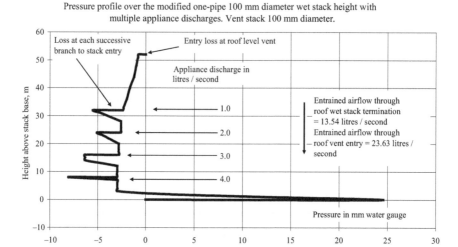

Figure 4.3 Comparable stack air pressure regime for a modified one-pipe network with both wet and dry stacks of 100 mm diameter

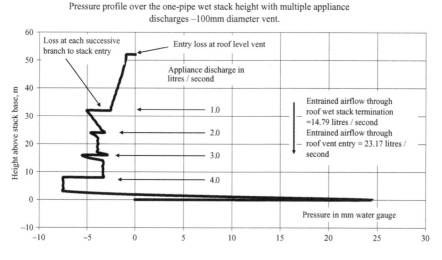

Figure 4.4 Comparable air pressure regime for a traditional one-pipe network with 100 mm wet and dry stacks

appear that providing the separate vent stack at the same diameter as the water carrying wet stack does limit the suction pressures attained in the network.

However design guidance available through various national codes indicates that the vent stack should be of a smaller diameter than the water carrying stack. The AIRNET simulation was employed to test this common assertion. Figure 4.5 illustrates the pressure regime when a 50 mm diameter vent stack is employed with a 100 mm diameter wet stack in a traditional one-pipe network.

In fact to improve the performance of the one-pipe system a larger diameter vent stack is required – clearly a proposition contrary to all current design codes but one substantiated by reference to the bifurcation of the pressure transient at a junction developed in earlier chapters. A larger diameter vent reduces the transmitted transient and hence reduces the stress placed on the appliance trap seals. However the extra material and space implications make this an impractical proposition. Figure 4.6 demonstrates the effect.

Distributing air admittance valves up the stack in a single stack network has advantages as it allows air to be drawn into the network closer to the point where it is needed to offset the negative transients that may deplete trap seals. Figures 4.7 and 4.8 demonstrate the effect of this design decision, in the first case by terminating the stack in an AAV and distributing further AAVs down the stack and secondly by retaining an open stack upper termination to atmosphere and distributing AAVs down the stack. Although the previously well-ordered pressure profiles are disrupted due to the natural vibration of the AAV diaphragm the negative pressure prevailing in the stack is again reduced.

The outcome of these comparisons is that the simulation confirms the earlier observational understanding of the shear force driven air entrainment in the

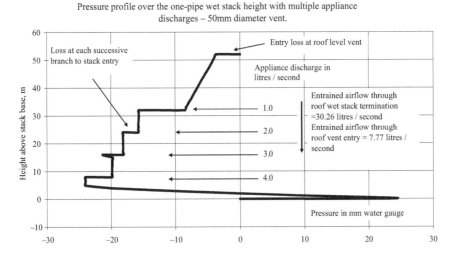

Figure 4.5 Pressure regime in the wet stack of a one-pipe network with 50 mm diameter vent, illustrating the degradation in performance compared with a 100 mm diameter vent

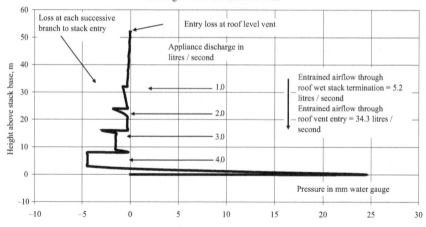

Pressure profile over the one-pipe wet stack height with multiple appliance discharges – 200 mm diameter vent.

Figure 4.6 Pressure regime in a one-pipe network with a 200 mm diameter vent stack in parallel with the 100 mm diameter water carrying stack

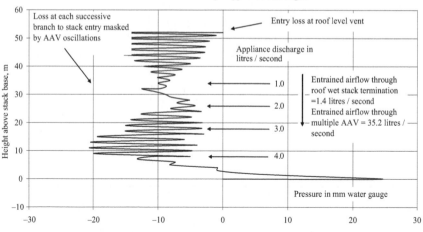

Pressure profile over the single-stack network with a stack top AAV and distributed AAVs with multiple appliance discharges.

Figure 4.7 Pressure regime in the single stack wet stack with AAVs fitted at the stack upper termination and at intervals down the stack

drainage network and the influence of passive venting. In addition the simulations presented provide an efficiency order for the various solutions considered. If the one-pipe system with a reduced 50 mm diameter vent parallel to the 100 mm diameter wet stack is taken as the reference it will be seen that the peak suction experienced is some 25 mm of water gauge below the lower discharging branch

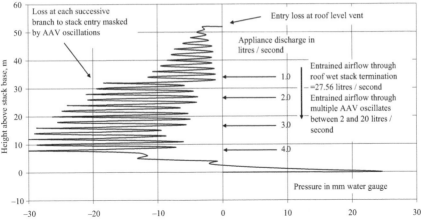

Pressure profile over the single-stack network and distributed AAVs and an open stack top termination with multiple appliance discharges.

Figure 4.8 Pressure regime in the single stack wet stack with AAVs fitted at intervals down the stack

connection. This is only matched by the single stack system with AAVs both distributed up the wet stack and used as the upper termination of the wet stack. The advantages in terms of material and space allocation are clear.

An alternative form of efficiency rating may be derived by considering the relative trap seal displacements and final trap seal retention for the systems simulated. Figure 4.9 illustrates the displacement of the trap seal water column in response to both positive and negative transients within the drainage network.

Negative pressure in the branch serving the appliance results in loss of trap seal water to the system, known as induced siphonage. It will be seen from Figure 4.9 that the maximum water column height on the system side remains zero as any negative pressure draws water into the branch. Conversely positive pressure displaces trap seal water upwards into the appliance bowl, whether a w.c. or basin/sink. As the appliance surface area increases rapidly above the entry to the trap, the simulation assumes a maximum upwards displacement, +90 mm water gauge being used in the simulations presented.

While the action of the system transients may wholly deplete the appliance trap, it is more likely that displacement of the water column will lead to a 'bubble through' condition where, once the water column on either the system or appliance side falls to the invert of the trap bend, an air path is formed that allows air to pass from the drain into habitable space or vice versa.

The AIRNET simulation is capable of predicting the displacement of the trap seal water columns and thus determining the retention of water seal at the culmination of the transient event. Figures 4.10 and 4.11 illustrate the form of these results for a single w.c. discharge to a vertical stack with and without a restricted upper vent termination system, as Figure 4.1(b).

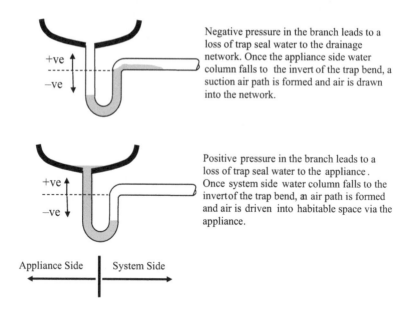

+ve
−ve

Negative pressure in the branch leads to a loss of trap seal water to the drainage network. Once the appliance side water column falls to the invert of the trap bend, a suction air path is formed and air is drawn into the network.

+ve
−ve

Positive pressure in the branch leads to a loss of trap seal water to the appliance. Once system side water column falls to the invert of the trap bend, an air path is formed and air is driven into habitable space via the appliance.

Appliance Side | System Side

Figure 4.9 Trap seal displacement in response to system transient propagation

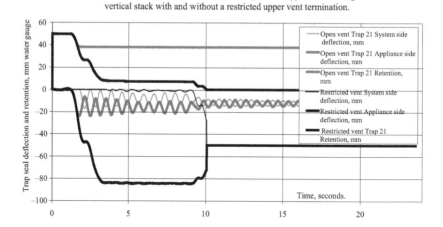

Trap seal retention and displacement following limited applaince discharge to a vertical stack with and without a restricted upper vent termination.

Trap seal deflection and retention, mm water gauge

60
40
20
0
−20
−40
−60
−80
−100

0   5   10   15   20

Time, seconds.

Open vent Trap 21 System side deflection, mm
Open vent Trap 21 Appliance side deflection, mm
Open vent Trap 21 Retention, mm
Restricted vent System side deflection, mm
Restricted vent Appliance side deflection, mm
Restricted vent Trap 21 Retention, mm

Figure 4.10 Trap seal retention and loss in response to a w.c. discharge to the vertical stack with and without a restricted upper vent termination; note trap 21 lost

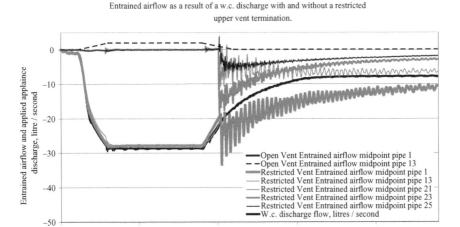

Figure 4.11 Entrained airflow in response to a w.c. discharge to the vertical stack with and without a restricted upper vent termination; note trap seal loss in the restricted vent case allows airflow through traps 21, 23 and 25

It will be seen from Figure 4.10 that restricting the upper termination leads to trap seal loss as the pressure in the stack falls due to the shear forces acting between the annular water film and the entrained air core. With an open upper termination the pressure reduction in the stack is mild and the trap seals are maintained.

Figure 4.11 illustrates the effect of trap seal loss as each of the lost trap seals now acts as an air inflow path – entrained airflow is shown through traps 21, 23 and 25 – namely the upper level traps. The simulation is therefore capable of modelling these events that arise as a result of the predicted air pressure regime within the system rather than as a result of some user-defined process. Figure 4.11 also demonstrates a fundamental prerequisite of any simulation, namely that flow at a junction sums to zero. This is demonstrated initially by the coincidence of pipe 1 and 13 airflows that show that total air entering the stack via the upper termination equates to the exit flow to the sewer, and later following loss of trap seal as the sum of inflows through pipes 13, 21, 23 and 25 equate to the exit airflow through pipe 1.

The simulation is also capable of modelling multiple discharges to the stack, as shown by Figures 4.2 to 4.9 at one particular time during the appliance discharges. Figure 4.12 presents the full test case where the development of an inflow profile from four floors is considered over a time period of 50 seconds. Figures 4.13 to 4.19 demonstrate the response of each of the systems already considered to this multiple increasing inflow, in each case both trap seal retention for all traps and the predicted water column oscillations for trap 19 are shown.

The outcome of these multiple inflow simulations, based on the trap seal loss not exceeding 25 per cent of the original 50 mm seal considered, i.e. a 37.5 mm retention, further confirms the earlier results at one flow rate at 38 seconds into

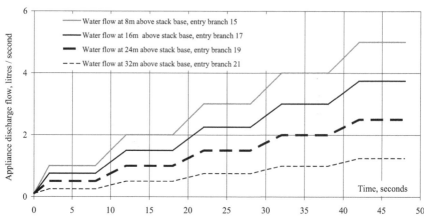

Figure 4.12 Increasing inflow from branches 15, 17, 19 and 21 over a 50 second period; the overall flow rises to 5 litres per second as shown

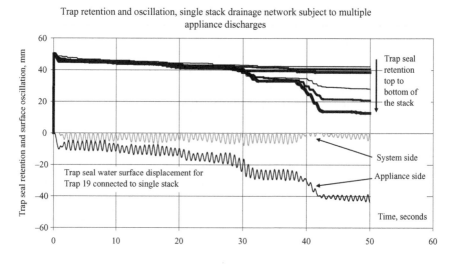

Figure 4.13 Trap seal retention for the single stack system together with the water column displacement for trap 19

the simulation. The 100 mm diameter single stack system is adequate at the lower water flows, however the 100 mm diameter one-pipe system with the normally recommended 50 mm diameter vent stack performs similarly, both being marginal at the maximum 5 litres/second discharge shown. In order for the traditional one-pipe system to outperform the other options the parallel vent stack has to be of a greater diameter than the wet stack – a design decision too far.

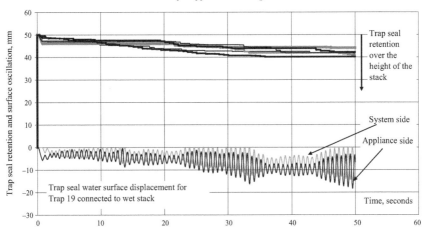

Figure 4.14 Trap seal retention for the modified one-pipe system together with the water column displacement for trap 19

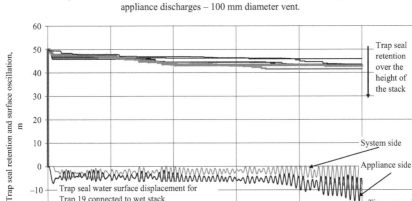

Figure 4.15 Trap seal retention for the one-pipe system with 100 mm vent stack together with the water column displacement for trap 19

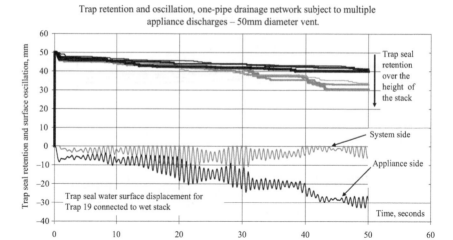

Figure 4.16 Trap seal retention for the one-pipe system with a 50 mm vent stack, as recommended in most codes, together with the water column displacement for trap 19

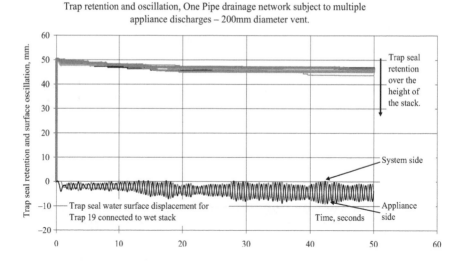

Figure 4.17 Trap seal retention for the one-pipe system with a 200 mm diameter vent stack, together with the water column displacement for trap 19

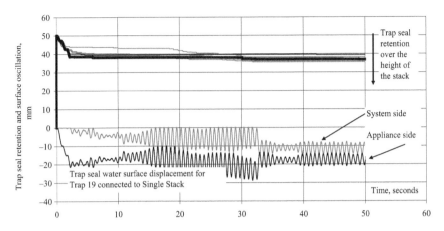

Figure 4.18 Trap seal retention for the single stack system with stack top and distributed AAVs, together with the water column displacement for trap 19

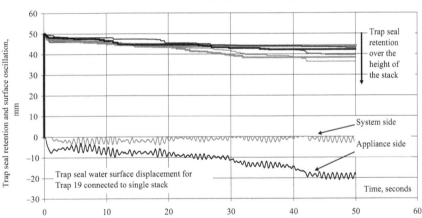

Figure 4.19 Trap seal retention for the single stack system with distributed AAVs, together with the water column displacement for trap 19

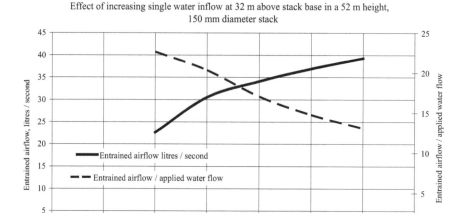

Figure 4.20 Increasing entrained airflow as a direct result of increased water inflow at one height, 32 m above stack base, for a 150 mm diameter single stack, Figure 4.1(b)

The modified one-pipe system – effectively a single stack with a parallel cross-connected vent – performs well when both stacks are 100 mm diameter.

It is noticeable that the distributed AAVs deliver an overall performance comparable to the modified one-pipe system with the obvious saving in installed material.

This section has therefore demonstrated that the simulation is capable of reproducing the expected pressure regime within the full range of drainage networks commonly utilised. The simulations presented have also demonstrated the ability to deal with multiple inflows as well as the simulation's ability to modify the boundary condition accessed dependent upon the local flow or trap seal retention condition.

4.2   Dependence of entrained airflow on appliance discharge, inflow position and restrictions to airflow entry due to blocked vent terminations or defective traps

As the applied water flow increases it would be expected that the entrained airflow would also increase, leading in turn to increased frictional and separation losses within the drainage network. To some extent this has been demonstrated by the previously discussed multiple inflow cases, however Figure 4.20 explicitly reproduces this effect for a 150 mm diameter single stack system, Figure 4.1(b), with the increasing inflow being applied at the 32 m height level through branch 21. As expected, the airflow increases with increasing water discharge, however the relationship will be particular to the system considered and it is not possible to propose a simple relationship as the balance set up between the shear forces entraining the airflow, and the frictional and separation losses dependent upon the particular design of the system, makes this an individual result that may only be predicted by means of the simulation techniques already discussed.

Similarly the wet stack height providing the base for the shear forces acting between the falling water annulus and the entrained air core must also influence the balance of forces determining the total entrained airflow experienced by any individual system. The simulation is capable of modelling this effect by simply moving the inflow branch, delivering 2 litres/second, up the stack, as shown in Figure 4.21. Again the result is particular to the system modelled, no simple relationship can be identified as again the final balance depends upon all the system parameters – for example the curvature of the branch entry to the vertical stack that determines the local loss at discharging junctions, or the degree of local restriction at the upper vent termination.

If the upper vent stack termination is wholly blocked and the traps connected to the system are assumed unable to allow an alternative airflow entry path, then the pressure in the stack will fall dramatically due to the shear forces that will still act between the annular water film and the stack air core. Figure 4.22 illustrates the pressure immediately below a blocked upper vent termination under these conditions; note that the pressure predicted will be oscillatory as the transients generated will be reflected between the stack upper termination, a dead end with a +1 reflection coefficient, and the stack base open entry to the sewer, a –1 reflection coefficient. In Figure 4.22, for ease of presentation only 2 second 'slices' of data are presented at each applied water flow rate. A best fit trend line has been added to Figure 4.22 to illustrate the falling mean pressure as the applied water flow increases to a maximum of 3 litres per second, again introduced at the 32 m height branch 21, Figure 4.1(b).

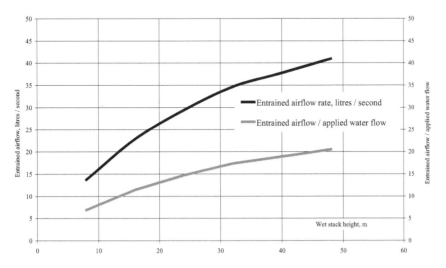

Figure 4.21 Increasing entrained airflow as a direct result of increased wet stack length by introducing a 2 litre/second water inflow further up the stack in a 150 mm diameter single stack system, Figure 4.1(b)

Clearly the pressure levels predicted in the blocked case would be sufficient to deplete any connected traps. Figure 4.23 illustrates the same test case where the connected traps have been allowed to fail and provide an alternative entrained airflow path into the network, demonstrating further the simulation's capacity to react to the conditions imposed on the drainage network.

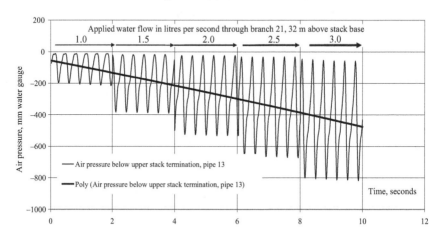

Figure 4.22 Pressure levels generated in a 150 mm diameter single stack where the upper vent termination and any depleted trap relief flowpaths have been blocked

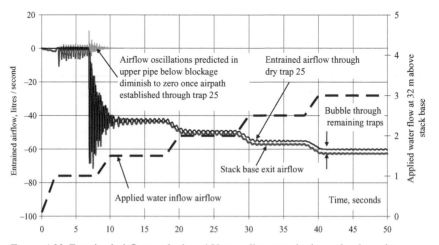

Figure 4.23 Entrained airflow paths in a 150 mm diameter single stack where the upper vent termination has been blocked

4.3 Simulation of transients imposed on the drainage network by external changes in conditions

The drainage networks illustrated in Figure 4.1 are all subject to potential changes in external conditions. These include wind shear effects over a roof level stack termination or transients propagated remotely in the larger sewer system, for example by pumping station start up or local surcharges. These transients will propagate into the building drainage and vent system and will affect trap seal retention and generate entrained airflows. It will also be necessary to be able to simulate local changes in conditions external to the network within the building itself, for example a trap within a space subject to an ambient negative pressure, possible through extract fan operation. These changes may be either time dependent or longer term.

Figure 4.24 illustrates the effect of a sinusoidal variation in external pressure over a vent stack roof termination. The sine wave, taken here as of amplitude 30 mm water gauge and 10 second period, propagates throughout the drainage and vent system, however the amplitude of the applied oscillation attenuates with distance from the vent termination. The oscillation period remains equal to the applied transient. Clearly such a boundary condition may be incorporated into any of the cases previously discussed as the simulation treats each boundary as a separate calculation so that a wind shear event illustrated here can be incorporated with any of the applied water flow cases dealt with previously or any of the sewer generated transients to be considered later in this section.

Transient pressure generated in the remote sewer system will propagate into the drainage and vent system and may lead to trap seal depletion and the ingress of foul air and sewer gas. The AIRNET simulation can model these transients by

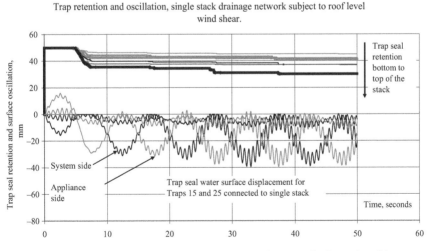

Figure 4.24 Pressure levels generated in a 150 mm diameter single stack subject to a sinusoidal wind shear, amplitude 30 mm water gauge, period 10 seconds, over the vent stack open roof termination

imposing a pressure regime at the stackbase boundary. Figure 4.25 illustrates a positive sewer pressure surge of 250 mm water gauge over 10 seconds imposed on the 150 mm diameter single stack system discussed above. During the sewer surge the single stack system continues to be subject to a 3 litre/second water inflow at the 32 m height point.

Figure 4.26 illustrates the effect on the system traps of the imposed sewer transient. System performance is satisfactory until the sewer system transient

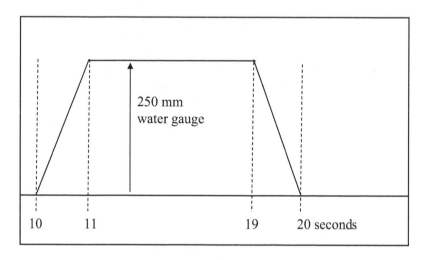

Figure 4.25 Imposed sewer transient at the stackbase boundary of a 150 mm diameter single stack system

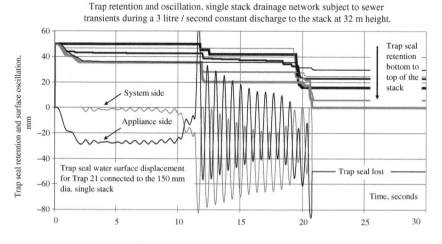

Figure 4.26 Trap seal response to a sewer transient imposed on a 150 mm diameter single stack system carrying a constant 3 litres/second water flow

arrives at 10 seconds, up to this time all trap seals are retained. The positive sewer transient displaces trap seal water into the appliance side of the traps and when the sewer transient is removed at 20 seconds trap 21 is lost as the displaced water column flows back through the trap and into the connected branch. Loss of trap 21 then allows an additional entrained airflow path into the system as illustrated by Figure 4.27.

These two cases, the roof wind shear and the sewer transient, demonstrate further the ability of the simulation to model a wide range of boundary conditions that may be user defined or arise as a result of the pressure regime developing within the network modelled. The imposition of a steady or time dependent 'room' pressure may similarly be dealt with as this merely alters the ambient conditions assumed to apply on the appliance side of the trap seal. This boundary condition will be returned to in a later treatment of the simulation of the 2003 Hong Kong SARS outbreak in Amoy Gardens where the presence of a bathroom fan reduced pressure on the appliance side of a dry trap and exacerbated airflow ingress from the drainage network.

4.4 The simulation of stackbase surcharge

The annular downflow in a drainage system vertical stack is transformed into free surface flow in the horizontal drain leading to the sewer connection at the stackbase. The transformation, Figure 4.28, establishes a water curtain at the base of the stack covering the entrance to the downstream drain at the higher water flowrates. The passage of entrained air through this water curtain results

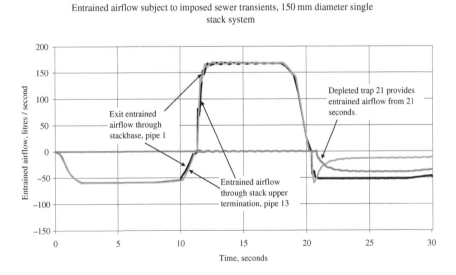

Figure 4.27 Entrained airflow through a 150 mm diameter single stack system exposed to a sewer transient while carrying a constant 3 litres/second water flow

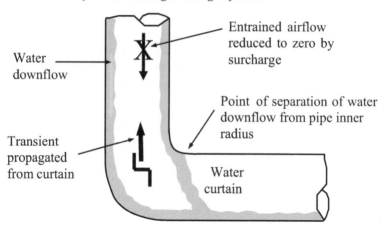

Figure 4.28 Water curtain at base of a stack demonstrating the possibility of a surcharge condition and a reduction in entrained airflow

in the characteristic positive back pressure experienced at the stackbase already discussed and included in the AIRNET simulation.

Under heavy flow conditions the thickness and resistance of the water curtain can increase substantially leading to a surcharge condition where the continued passage of the entrained airflow is severely curtailed and in some cases the airflow may be reduced to near zero. While the simplest model would reduce airflow to zero under surcharge conditions, this is not a realistic model as the increasing pressure in the stack will force an airflow through the water curtain. The model used in AIRNET is that the curtain resistance is allowed to increase rapidly and this then generates the positive surge that characterises surcharge conditions. The increase in curtain resistance is achieved by imposing a notional enhanced water flow at the stackbase thus allowing the model curtain resistance to rise.

Based on the Joukowsky expression for a reduction in fluid velocity in less than one pipe period it will be appreciated that for air, density 1.3 kg/m³ and wave speed 320 m/s, a reduction in airflow velocity of 1 m/s will generate a positive surge of 40 mm water gauge. Thus the surcharge condition may be responsible for trap seal depletions due to overpressure, as demonstrated in Figure 4.26 for a sewer positive transient.

As shown in Figure 4.29 the effect of the surcharge between 5 and 6 seconds into the simulation is to generate a positive transient that displaces the trap water column into the adjacent appliances. As the surcharge abates at 6 seconds, the reduction in stack base blockage allows a resumption of entrained airflow and the stack pressure reduces allowing the trap seal water to pass back into the traps and trap 15 is lost as the column has sufficient momentum to self-siphon. Entrained airflow then enters the system through the depleted trap 15.

While it is attractive to view the stack base surcharge as a simple increase in water curtain resistance, simulated by introducing an enhanced notional water flow at the stack base, in practice the process is much less well ordered. It has

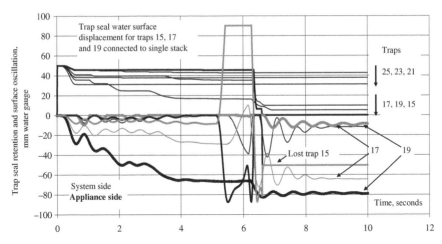

Figure 4.29 Trap seal retention and water column displacement in a 100 mm single stack system subjected to a stack base surcharge

been observed that the annular flow in the vertical stack displays non-uniform flow conditions during notionally steady flow as the annular film carries roll waves, discussed in Chapter 1, Filsell (2006). If the stack base boundary is represented by an oscillatory resistance, initial results suggested a remarkably constant 4 Hz oscillation, modelling the arrival of a series of roll waves that impose a time dependency on the stack base water curtain resistance, then it is possible to simulate the propagation of pressure transients at the roll wave arrival frequency. Filsell also identified the oscillatory nature of the water curtain formed at a discharging branch to stack junction; initial results suggested a 12 Hz oscillation, and it is therefore also possible to modify the junction boundary condition to include a notional oscillation in junction resistance. Introducing both these modified boundary conditions allows the oscillatory nature of entrained airflow to be modelled, Jack, Swaffield and Filsell (2004), Figure 4.30. Clearly this is a simulation based on initial laboratory measurement of the water curtain oscillations, however it illustrates the mechanisms involved and suggests areas for further investigation, as well as confirming the ability of the AIRNET model to deal with user defined and time dependent boundary conditions.

   In order to reduce the effect of the stack base surcharge event (Figure 4.29) it is possible to simulate the presence of a positive transient attenuator (PTA) at the base of the stack and on branches further up the stack. A series of PTA units were simulated – 16 litre maximum on pipe 26 with 10 litre maximum units on pipes 27, 28, 29, 30 and 31, Figure 4.1(b). Figure 4.31 illustrates the modification to the trap seal retention and water column displacements as a result of this installation. The effect of this introduction of a diversion route for the entrained airflow restricted by the stack base surcharge is a marked reduction in the peak pressures and the

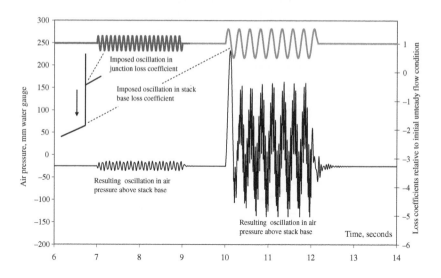

Figure 4.30 Response of the stack air pressure to time dependent oscillations in both junction and stack base loss coefficients

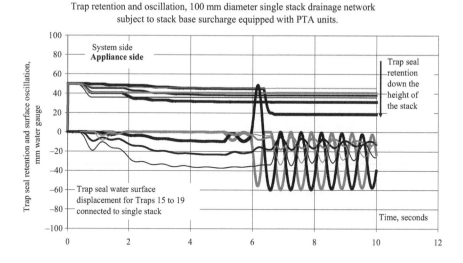

Figure 4.31 PTA installation simulated on pipes 26 to 31, initial 16 litre PTA on pipe 26 with maximum 10 litre units on pipes 27, 28, 29, 30 and 31

retention of all trap seals when compared with the unmodified system, Figure 4.29. However the trap seal retention for trap 15 is less than the 37.5 mm water gauge recommended in design codes.

Figure 4.32 illustrates the entrained airflows predicted within the network in response to the stack base surcharge. It will be seen that the airflow exiting to the sewer is curtailed during the surcharge event and that this reduction in throughflow is balanced by an airflow into the lowest level PTA unit on pipe 26 until the PTA becomes pressurised. The next PTA on the system is connected to pipe 27 and Figure 4.32 illustrates an inflow to this unit as soon as the PTA on pipe 26 is pressurised. In this simulation the higher level PTA units are not required and remain deflated in response to the ambient negative pressures in the stack.

Figure 4.33 illustrates the inflation of the PTA units in response to the surcharge. It is stressed that the deflated PTA offers a diversion path for the entrained airflow that would otherwise have been retarded by the enhanced water curtain effect during surcharge. Once the PTA is fully inflated to its nominal atmospheric pressure volume it will pressurise and can be of no further assistance in attenuating the effect of the surcharge. The positive pressure wave therefore passes on up the stack to affect the next trap seal or be attenuated by the next PTA unit. This effect is demonstrated by the secondary inflation of the pipe 27 PTA that is seen to commence as the lower level pipe 26 PTA reaches its 'full' inflation of 16 litres at atmospheric pressure. The simulation therefore is shown to be capable of applying the correct choice of boundary condition dependent upon the local pressure conditions.

In the simulations undertaken, the next logical step was to increase the volume of the pipe 26 PTA and this was set to 30 litres. Figure 4.34 simulates the effect of such an installation – the equivalent of 8 PTA units over the lower floors served

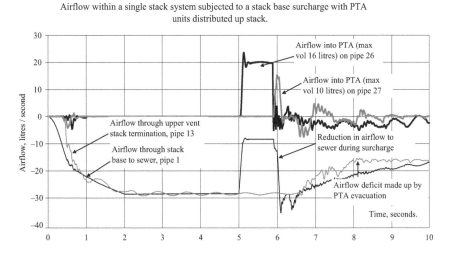

Airflow within a single stack system subjected to a stack base surcharge with PTA units distributed up stack.

Figure 4.32 Entrained airflow predicted within a single stack system protected by the installation of PTA units

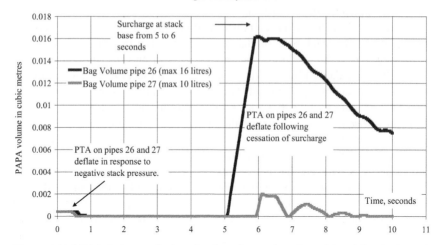

Multiple PTA operation in response to a stack base surcharge in a 100 mm diameter
single stack system

Figure 4.33 PTA inflation predictions for the units on pipes 26 and 27 of the single stack
system in response to a stack base surcharge from 5 to 6 seconds into the simulation

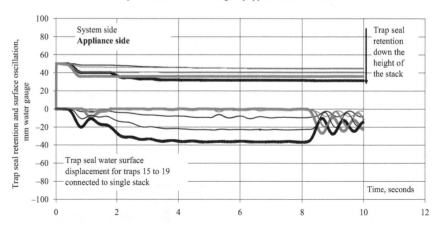

Trap retention and oscillation, 100 mm diameter single stack drainage network
subject to stack base surcharge equipped with PTA units.

Figure 4.34 Trap seal retention and water column displacement in a 100 mm single stack
system fitted with PTA units to suppress the stack base surcharge

by the stack. In this case the trap seal retention is satisfactory with no positive
pressure being recorded at the traps above the pipe 26 PTA installation.

Therefore a satisfactory, if not necessarily optimum, PTA volume was
determined by successive simulations to suppress the positive surge pressures
responsible for the trap seal depletion demonstrated by Figure 4.29. It will be

shown later that early P.A.P.A.™ installations were determined on site by this experiential process of adding volume until the surcharge problem was removed – a successful example of this technique from Hong Kong will be discussed in Chapter 5.

These simulations demonstrate the importance of the 'rules of surge suppression' – namely that the surge protection must be positioned between the source of the transient and the equipment to be protected and that the fundamental concept of surge protection is to identify means of reducing the rate of change of the fluid flow by providing an alternative flow path and a means of gradual deceleration – in this case provided by the inflow into the PTA and the gradually rising resistance to flow provided by its pressurisation.

It will be seen in Figure 4.34 that there is no positive trap seal displacement on trap 15 and this is explained as the PTA has been positioned between the source of the transient – the stack base surcharge – and the branch leading to trap 15. The volume of the PTA is sufficient to wholly suppress the positive transient, thereby maintaining trap seal integrity.

4.5   Modelling the effect of a surcharge in a stack offset

The design of some building types dictates that offsets in a vertical stack cannot be avoided, examples would be high-rise developments featuring a podium common lower floor element where multiple vertical stacks are brought together prior to connection to a sewer. Generally offsets are to be avoided as they merely provide a series of additional stack base surcharge opportunities. Early drainage designers introduced offsets as they believed that this would 'slow down' the annular water flow in the stack – this is still heard sometimes and ignores the terminal velocity of the annular flow now generally accepted.

Figure 4.35 illustrates a surcharged offset and confirms that a transient is propagated in both the upstream and downstream stacks – the upstream transient being positive as the airflow is reduced while for the same reason the downstream transient is negative. In the worse case of an instantaneous cessation of airflow both will have the Joukowsky instantaneous flow stoppage value of 41 mm water gauge per m/s of airflow destroyed, see Chapter 3.

The offset therefore forms a downstream boundary for the upstream section of wet stack and an upstream boundary for the stack below the offset; in both cases the boundary may be represented by putting the local airflow velocity to zero for as long as the surcharge persists. In practice the cessation of airflow is not likely to be instantaneous or total as the rising air pressure forces airflow through the surcharge which itself is composed of a water mass that continues to move down the stack. However these boundary conditions allow a consideration of mitigation design to limit the surcharge effect.

Generally national codes suggest bypass venting to alleviate the pressure surge experienced in the stack. An alternative would be to introduce Active Control by relieving the transient upstream of the offset with a PTA and below by introducing an AAV; Figure 4.36 illustrates these options.

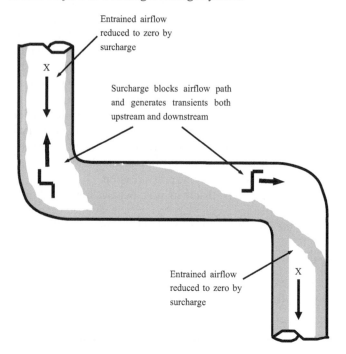

Figure 4.35 Offset surcharge in a vertical stack

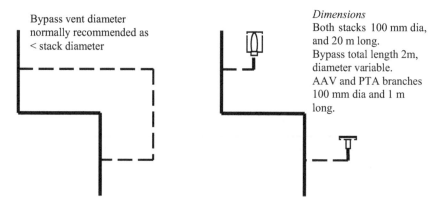

Figure 4.36 Offset venting or installation of an AAV / PTA combination

The effect of bypass vent diameter may be investigated by applying the simulation techniques developed in Chapter 3 and demonstrated in this chapter. It will be assumed that the airflow stoppage is instantaneous and complete.

Figure 4.37 illustrates the effect of bypass diameter choice on the distribution of entrained airflow around the offset. As the bypass increases in diameter from

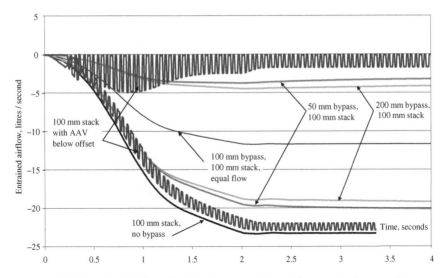

Entrained airflow through main stack, bypass and AAV when included.

Figure 4.37 Entrained airflow distribution around the offset as the bypass diameter increases; note the entrained airflow through the AAV mounted below the offset

50 to 200 mm so the percentage of airflow bypassing the offset increases. Note that when the bypass and stack diameters are equal, the airflow splits equally along each route prior to the surcharge event. Commonly national codes suggest a bypass diameter less than the stack diameter.

The inclusion of an AAV below the offset introduces an additional small entrained airflow prior to the surcharge as shown in Figure 4.37.

A surcharge event at the offset may be caused by a sudden increase in the stack downflow or the passage of solids through the systems. Figure 4.38 illustrates the pressure predictions immediately above and below the offset with no bypass or Active Control protection installed.

Figure 4.38 demonstrates the influence of boundary reflection coefficients. The upper stack termination may be represented by a −1 reflection coefficient as the stack terminates to atmospheric pressure so that the pressure trace during the surcharge displays alternative positive and negative transient propagation so that the oscillation is about the initial upstream line pressure.

Downstream of the offset the stoppage of the entrained airflow generates an initial negative transient equal to the positive propagated above the offset. However the situation below the offset is complicated by the continuing water flow down the stack. There continues to be a shear force driven suction applied below the offset and this is represented by the continuing fall in downstream pressure predicted by the simulation, demonstrated by Figure 4.39, so that the pressure transient reflections are superimposed on this continuing downward pressure impetus.

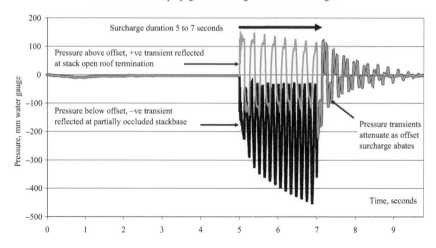

Figure 4.38 Pressure above and below the offset with no surge protection

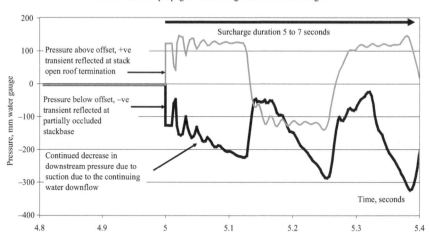

Figure 4.39 Continuing water downflow further reduces the pressure below the offset during surcharge

The effect of each of the bypass configurations considered, namely 50–200 mm diameter bypass and various Active Control PTA and AAV combinations may also be simulated. Figure 4.40 demonstrates the effect over the first two pipe periods following surcharge.

As shown by Figure 4.40 the efficiency of the offset increases with the diameter of the bypass. Generally codes recommend bypass and parallel venting of a smaller diameter than the wet stack, however this may be shown to be flawed, a result supported by these simulations. The Active Control installation of a 12 litre PTA and AAV combination is more efficient than the 50 mm diameter bypass and 100 mm stack combination as shown by Figures 4.40 and 4.41. It will be seen from Figure 4.41 that the PTA bag inflates at atmospheric pressure until the 12 litre limit is reached and then the bag pressurises. Specifying the initial PTA volume may be undertaken by use of the simulation presented herein. In practice the applied surcharge will not be instantaneous so that the reduction in transient pressures experienced once the change in flow velocity takes longer than one pipe period, discussed in Chapter 2, will alleviate the applied transients and allow the use of smaller PTA volumes.

Figure 4.42 demonstrates the effect of varying the PTA volume from 6 to 24 litres, a result that may be achieved by parallel or series mounting of the variable volume containment device, as will be demonstrated in an actual application in Chapter 5. As the bag volume increases, the PTA is capable of absorbing the transients and does not demonstrate the cyclic pressurisation/depressurisation curve, cf. 24 and 6 litre results in Figure 4.41.

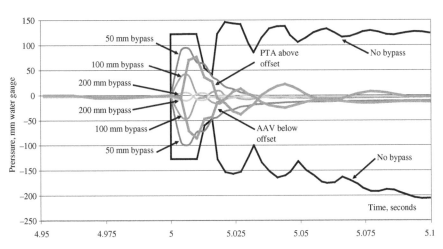

Pressure levels above and below bypass and active control connections.

Figure 4.40 Initial pressure surge following surcharge of offset as mitigated by bypass and Active Control installations

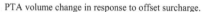

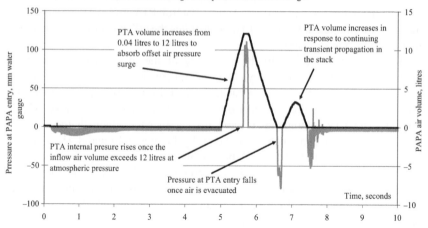

Figure 4.41 Pressure mitigation above the offset due to the introduction of a 12 litre volume PTA

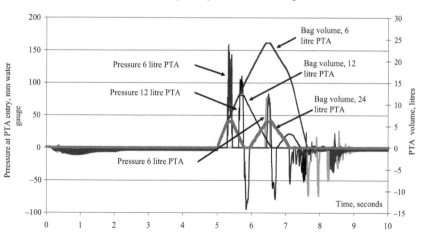

Figure 4.42 PTA volume determines the degree of pressure alleviation achieved upstream of the offset

Offsets are not to be recommended, however building design often allows no option. In such cases bypass or Active Control options can be used to limit the pressure transient effects of surcharge.

4.6   Frictional representation as affected by the rate of change of the local flow conditions

The representation of friction within the Method of Characteristics numerical solution of the St Venant equations is normally restricted to the use of a quasi-steady approach where the local velocity at the start of a computing time step is used to generate a frictional term within both characteristic equations. This 'first order' approximation has historically been found to be satisfactory, Streeter and Lai (1962), however a second order approximation was also used where the friction term was based on the average of the initial and final velocities across a computing time step, but the added complication of introducing the square of unknown final velocity was not found to be justified.

From Chapter 3 the C^+ characteristics linking known conditions at time t at an adjacent upstream node R to the unknown conditions at $t+\Delta t$ at the local node P is

$$u_P - u_R + \frac{2}{\gamma - 1}\left(c_P - c_R\right) + 4 f_R u_R \mid u_R \mid \frac{\Delta t}{2D} = 0$$

while the C^- characteristics linking known conditions at time t at an adjacent downstream node S to the unknown conditions at $t+\Delta t$ at the local node P is

$$u_P - u_S - \frac{2}{\gamma - 1}\left(c_P - c_S\right) + 4 f_S u_S \mid u_S \mid \frac{\Delta t}{2D} = 0$$

However the quasi-steady approach is actually flawed as it assumes that the flow remains steady and uniform whereas in a transient situation, as described in Chapter 3, there may be local reversals of flow across a conduit and therefore the assumption that a steady uniform model of frictional resistance involving for example the Colebrook–White friction factor relationship is sufficient is not wholly justified. Early efforts to improve the frictional representation included the addition of a term based on the local time acceleration, Carstens and Roller (1959).

Nevertheless the quasi-steady model is universally used within surge analysis using MoC for the pragmatic reason that the rate of change of the flow conditions are sufficiently slow, in terms of time and pipe period, for the predictions to remain valid within normal engineering limits – i.e. the other parameters within the system are not known to any higher degree of accuracy.

Zilke (1968) provided the seminal work on unsteady friction in laminar flows, discussed in Chapter 3, showing that for laminar flow it was possible to predict the effect of flow disruption during the passage of a transient on the frictional resistance term necessary within the MoC solution. Effectively Zilke's work allowed the time history of the local flow velocity to be included as a summable series within the calculation of the friction term, having the overall effect of increasing the friction. Zilke's work has since been extended to turbulent flows and remains a regular topic of advanced papers in surge analysis, Vitkovsky et al. (2004).

It is useful to consider whether there is any advantage in moving to a more computationally rigorous approach for the modelling of low amplitude air pressure transients in building drainage and vent systems, Swaffield (2007). It will be assumed that it has already been shown in the surge literature that there is no advantage in moving to a second order approximation so the choice is between the first order simplified model currently used where Colebrook–White is conventionally employed to yield the friction factor term for the characteristic equations or moving to some representation of unsteady friction. However it is instructive to consider a comparison between the quasi-steady approach and a combination of the Zilke solution for friction within the laminar flow conditions found in an appliance trap and the Carstens model for friction factor in the drainage system airflows. The variable that will determine the boundary of application for the current methodology is the rate of change of flow conditions brought about by the transients propagated throughout the network.

A series of test cases will be discussed, each based on a simple network as illustrated in Figure 4.43. The transient response of the network to a rising rate of change of water discharge will be simulated using both the traditional quasi-steady frictional representation involving Colebrook–White and an unsteady representation involving both the Zilke and Carstens models as already discussed. A water discharge of 2 litres per second will be taken as the test case with rise times varying from 5 to 0.1 seconds. Deceleration will be at the same rate and the steady flow will be maintained within an overall event time of 15 seconds. In order to avoid the complication of trap seal loss it will be seen from Figure 4.43 that the 100 mm deep trap will only have 40 mm of water depth in each of the system and appliance side columns at the start of the simulation to avoid inadvertent water loss.

Figure 4.43 presents the test network and Figure 4.44 the applied water discharge profiles.

The results of the investigation are best presented as a series of comparisons at different rise times, from 5 seconds down to 0.1 seconds; Figures 4.45 to 4.48 presents the initial comparison between the 5 and 2 second rise times for both quasi-steady and unsteady frictional representations. As will be seen there is no discernable advantage to using the more complex unsteady formulation for frictional resistance. This result is replicated in Figure 4.46 where there is little difference between the two approaches for rise times from 1 second to 5 seconds.

However at 0.5 seconds it is apparent that the quasi-steady frictional model tends to overestimate the oscillation of the water column, Figure 4.47. This is understandable as the Zilke model applied to the friction between the trap seal water and the trap wall surfaces will predict a higher friction factor and will therefore attenuate any column oscillation.

This effect is exacerbated in the 0.1 second rise time case, Figure 4.48, as here the quasi-steady model predicts the loss of the trap seal water due to immediate induced siphonage into the system. The Zilke model again predicts a greater

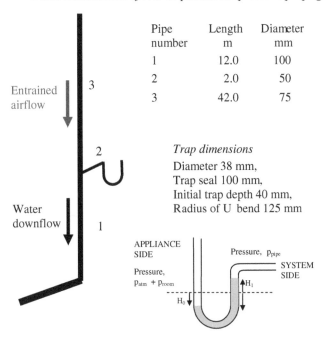

Pipe number	Length m	Diameter mm
1	12.0	100
2	2.0	50
3	42.0	75

Trap dimensions

Diameter 38 mm,
Trap seal 100 mm,
Initial trap depth 40 mm,
Radius of U bend 125 mm

Figure 4.43 Single stack network used in a simulation assessment of the necessity to include unsteady friction

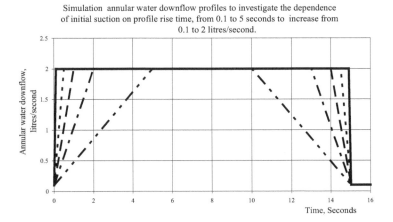

Figure 4.44 Applied water discharge profiles to investigate the limits of quasi-steady frictional modelling in the case of low amplitude air pressure transient propagation in building drainage systems

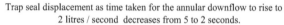

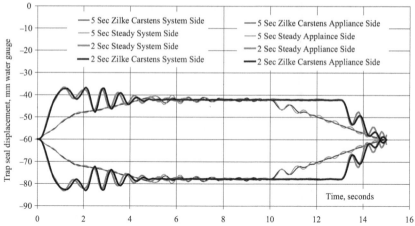

Figure 4.45 Comparison of the frictional representations for rise times from 2 to 5 seconds – no advantage in an unsteady formulation

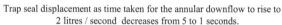

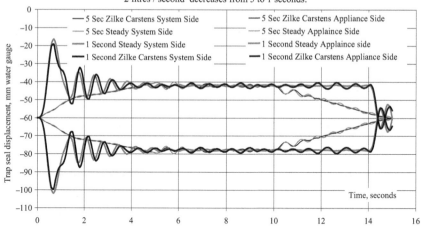

Figure 4.46 Comparison of the frictional representations for rise times from 1 to 5 seconds – no advantage in an unsteady formulation

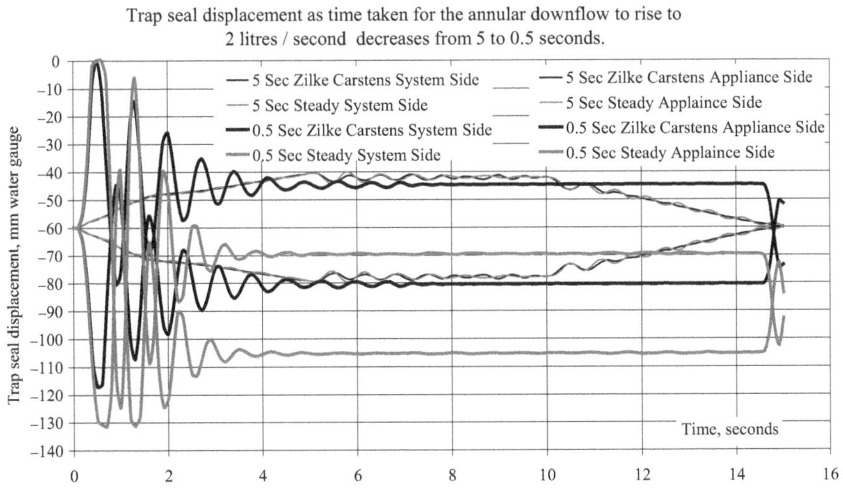

Trap seal displacement as time taken for the annular downflow to rise to
2 litres / second decreases from 5 to 0.5 seconds.

Figure 4.47 Comparison of the frictional representations for rise times from 0.5 to 5 seconds – indication that the quasi-steady approach enhances the predicted oscillation during the initial profile rise time

Trap seal displacement as time taken for the annular downflow to rise to
2 litres / second decreases from 5 to 0.1 seconds.

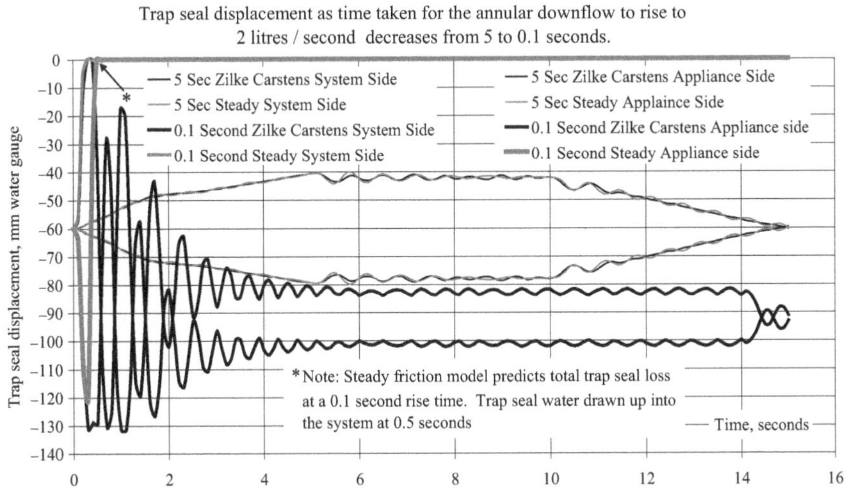

Figure 4.48 Comparison of the frictional representations for rise times from 0.1 to 5 seconds – the quasi-steady approach enhances the predicted oscillation during the initial profile rise time and predicts a total trap seal loss

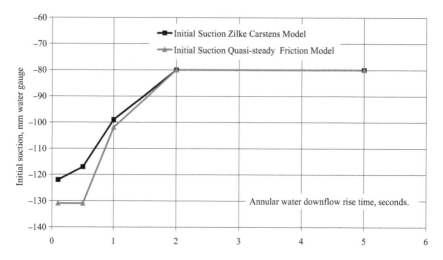

Figure 4.49 Effect of annular water downflow rise times on the initial suction predicted at the appliance trap seal

friction term, particularly as the rate of change of the flow conditions is now very rapid, and predicts a retained trap seal.

In order to evaluate the significance of these results, summarised by Figure 4.49, it is necessary to consider the likely rates of change of flow conditions within the systems to be simulated. The rate of change of flow conditions within a building drainage system will depend upon both the applied water inflows and upon any transient generating effects caused by flow interactions with the system.

The majority of appliance discharges to the network will be relatively gradual in terms of the pipe periods expected within the network. The range of rise times considered in this study was from 0.4 litres/second/second to 20 litres/second/second. If the w.c. is considered as the most likely appliance to feature a rapid rise time then this would suggest a rise to 2 litres/second in 2 seconds as a typical 'fast' rise time. Appliances such as baths, sinks and showers will have lower peak flows and longer rise times. It is likely that pumped waste from dishwashers and washing machines will also be gradual.

Transients will also be generated by interactions within the system and here the most obvious candidate for concern is the stack base surcharge event, however again it is likely that the rise time would exceed the 0.1 seconds seen in the study as problematic. Similarly wind effects over open roof vents may introduce repeating waveforms, however the wind generated frequency would be lower than required to introduce problems.

One example where the effect is essential will be returned to in more detail in Chapter 6, namely the development of non-invasive technique to identify dry trap seals that is based on the introduction into the system of a 10 Hz sinusoidal pressure wave with peak amplitudes under 30 mm water gauge. Here the use of the Zilke Carstens model is essential to predict the trap seal response. The quasi-steady model will predict a trap seal oscillation under these conditions roughly 10 times that predicted by the unsteady model that also replicates observed trap seal oscillations.

4.7 Discharging branch boundary condition and the effect of falling solids

Chapter 3 introduced the local separation losses due to the passage of the entrained airflow through the water curtain established across a wet stack junction by a discharging branch, while Figure 1.2 illustrates the varying water curtain coverage. It will be appreciated from Figure 1.2 that the loss coefficient applicable at the discharging branch to stack junction depends upon the relative diameter ratio of the branch and stack, the ratio of branch flow to total stack flow below the junction and the radius of curvature of the branch inlet:

$$\Delta p_{junction} = \frac{1}{2} \rho K_{loss(Q_{stack},Q_{branch},stackgeometry)} u^2_{stack} \tag{4.1}$$

$$K_{loss} = \phi \left[\frac{D_{branch}}{D_{wetstack}}, \frac{Q_{branch}}{Q_{wetstack}}, R_{entry} \right] \tag{4.2}$$

The effect of branch inlet radius of curvature will be demonstrated in Chapter 8 in a discussion of maximum allowable stack flows. O'Sullivan (1974) investigated the value of the junction loss coefficient for a range of branch to stack ratios, a range of inlet radii of curvature and flow ratios, including the case of the first discharging branch where Q_{branch}/Q_{stack} will be unity, these values being incorporated into AIRNET, Jack (1997).

The passage of a falling solid down the vertical stack following insertion at an upper branch will generate transient propagation above and below any discharging branch encountered. Gormley (2007a) investigated this transient propagation during a series of site tests in a 17-storey building. The pressure regime within the stack was monitored over its height by mounting transducers above each junction in the vertical stack. The effect of the solid passing through each discharging junction water curtain was clear, Figure 4.50, the disruption to the water curtain propagates a positive transient downwards and a negative transient upwards as the entrained airflow demand rises. Closure of the curtain following passage of the solid reverses this mechanism so that both transients appear as pulses of short duration.

The effect of the solid passing through the water curtain was to instantaneously and for a short duration modify the curtain loss coefficient. Gormley (2007a) simulated this effect by reducing the junction loss coefficient in the AIRNET simulation for a short duration. The resulting simulation, Figure 4.51 (a) and (b), demonstrates the same effect as shown by the site tests, Figure 4.50. Generally the

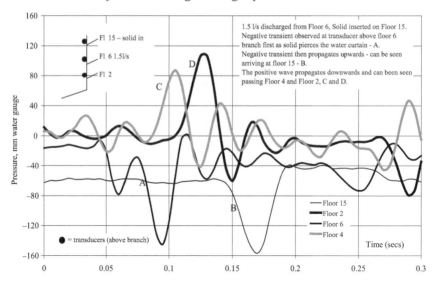

Figure 4.50 Identification of the positive (downstream) and negative (upstream) transients generated as the falling solid traverses a discharging junction water curtain

duration of the transients associated with falling solids are of too short a duration to realistically affect trap seal retention.

Gormley (2007a) also used the pressure traces illustrated to estimate both the acoustic velocity in air as the propagated transient passes each transducer station, 320 m/s being recorded, and the falling solid velocity by introducing a second discharging junction, in the case discussed 4.27 m/s over a 12 m fall; however it is clear that further experimental investigations would be necessary to determine the solid terminal velocity based on its mass and the drag forces acting within stacks of various diameters. The falling solid could be included in the AIRNET simulation as a moving boundary condition that would indicate the times during the simulation when any discharging branch water curtain loss coefficient should be reduced, however the work reported does not indicate that this would be required. Moving boundary conditions will be returned to in Chapter 7 where the necessary modifications to the MoC solution will be discussed in the context of train and elevator transients.

4.8   Concluding remarks

This chapter had as its objective the demonstration of the simulation techniques introduced in Chapter 3 and in particular the application of the wide range of boundary conditions seen to represent the operation of a building drainage and vent system subjected to low amplitude air pressure transient propagation. This chapter has therefore by definition been mainly theoretical as it has used convenient test systems to draw out the various issues surrounding transient propagation. In addition

(a)

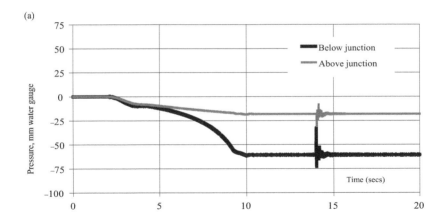

(b)

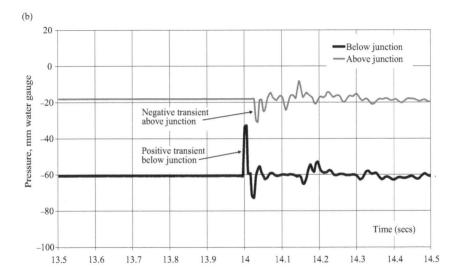

Figure 4.51 (a) and (b) Simulation of the effect of a falling solid, confirming the propagation of positive (downstream) and negative (upstream) transients as the discharging junction water curtain is disrupted

the chapter has attempted to provide a systematic approach to transient propagation in drainage networks while at the same time building a base of understanding that will allow later practical applications that may involve simultaneously more than one of the cases introduced here.

5 Pressure surge as a source of system failure, leading to the development of control and suppression strategies

Earlier chapters have stressed that the propagation of pressure transients within fluid carrying systems is a natural consequence of any intentional or inadvertent change in the system operating condition and is the mechanism by which change is communicated to the system as a whole. Due to the universal nature of pressure transient propagation, its analysis, simulation and, most importantly, its prevention and control, became a truly international interdisciplinary engineering topic whose genesis has already been discussed, from the seminal work of Joukowsky that, in 1900, established the fundamental understanding of transient propagation, through to the introduction of computer based Method of Characteristics analysis techniques in the 1960s, due to the work of Lister, Streeter and Fox.

Two interlinked issues dominated Joukowsky's study, namely the surge pressures generated by a change in flow conditions and the celerity at which these pressure waves were transmitted through the network. A further issue, clarified by Joukowsky, was the interaction between an arriving pressure wave and the local physical nature of the system – referred to in more modern studies as the boundary conditions for the network. However the fundamental drive behind this original work, and indeed a continuing central theme in transient analysis, is the prevention, suppression and control of transients. Pressure surge, as will be discussed, may lead to catastrophic system failure, as evidenced by the massive pressure surges capable of destroying pipelines and control structures, or low amplitude air pressure transients sufficient to deplete appliance trap seals and allow cross-contamination of habitable space – both extremes are examples of system failure due to the propagation of pressure transients and have an equal capacity to cause fatalities.

5.1 Consequence of transient propagation

Changes in operating point lead to pressure transient propagation where the severity of the transient has already been defined as dependent upon the fluid (its density and bulk modulus and entrained gas content if a liquid), the properties of the conduit wall (its Young's modulus and diameter to thickness ratio), and most importantly the rate of change of the flow condition, represented most simply by the rate and magnitude of the incremental velocity change. It has already been

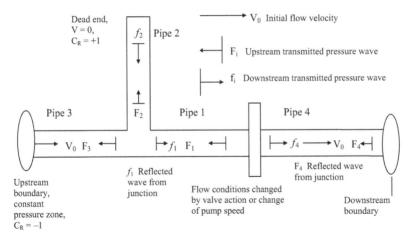

Figure 5.1 Pressure transient propagation, illustrating the concept of wave reflection and transmission at naturally occurring boundaries

stressed that the relevant time measurement is the pipe period, rather than absolute time, as it is the arrival of pressure wave reflections from the network boundaries that defines the overall pressure response at any location. This concept will be particularly important in the later treatment of traditional passive approaches to pressure transient control in building drainage and vent systems.

Reductions in flow velocity, due to a valve closure or pump failure, generate positive transients that travel upstream and negative transients that propagate downstream. Thus the simplest system failure would involve a rupture of the conduit upstream of a valve following rapid closure – defined as completion of closure in less than one pipe period, Figure 5.1. This form of failure is relatively straightforward as it is clear where and when the peak pressure will occur. Conversely the rapid opening of a valve that isolated the system from a high pressure source would result in the propagation of a positive pressure wave downstream that accelerates the flow.

More hazardous consequences arise as a result of interactions between the transient and the network. Rapid closure of a valve or the failure of a pump will propagate a negative pressure wave downstream that results in the formation of local fluid vapour pockets downstream, or the release of dissolved gas. In this case column separation occurs and the separated fluid columns upstream and downstream of the cavity move independently, the subsequent flow–time history being driven by the prevailing pressure gradients in the network. The rejoining of the column may generate pressure surge levels in excess of those predicted by a consideration of the initial change in system operating point, Figure 5.2, Swaffield (1970, 1972a). Alternatively the negative transient may lead to conduit implosion as the local pressures may fall to fluid vapour pressure. The presence of dissolved gas released from the fluid as its pressure falls below atmospheric has the effect of maintaining the system pressure above vapour level, Figure 5.2, however the resulting compression and decompression of the free gas pocket results in a slowly attenuating pressure

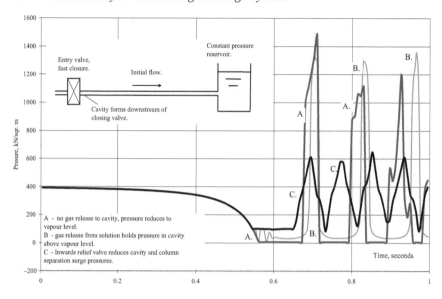

Figure 5.2 Pressure surge following a rapid valve closure, indicating the onset of column separation and the resulting pressures on cavity collapse or released gas compression

surge. The addition of an inwards relief valve attenuates the collapse pressure surge, again illustrated in Figure 5.2. In these cases the upstream boundary conditions are provided initially by the valve discharge loss coefficient, then by the fluid vapour pressure or the application of the gas laws to gas released from solution as the fluid pressure falls. The duration of application of these boundary conditions is controlled by the growth and collapse of a theoretical cavity at the valve location, Doyle and Swaffield (1972). Once the cavity collapses the boundary condition is provided by zero flow velocity, until the complex reflections in the system cause the pressure to again fall to vapour pressure. The presence of an inwards relief valve is simulated by a boundary condition that includes the inflow loss coefficient. Gas once released from solution is often assumed to remain free and hence forms a boundary condition controlled by the gas laws.

The presence of trapped air adds a serious complication to the process of transient propagation. An incoming positive transient approaching a dead-ended conduit with a trapped gas volume present will initially compress the trapped air with little or no effect on the column that will continue to accelerate until the compressed gas reaches a high enough pressure to cause rapid deceleration, Figure 5.3. The positive transients propagated as a result may exceed by an order of magnitude the initial column driving pressure, Martin (1976). Instances where this effect has led to catastrophic failure within building services systems include the failure of air release valves on multi-storey building dry riser installations and fire-fighting sprinkler systems dry prior to activation, Lawson, O'Neill and Graze (1963) and Hope and Papworth (1980), and the explosive failure of a

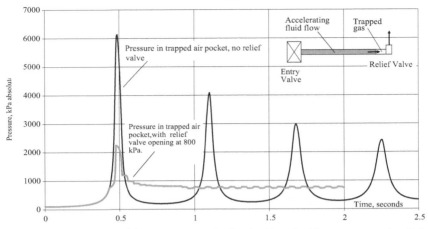

Figure 5.3 Comparison of the pressure at a dead end with and without an outwards relief valve following a rapid valve controlled inflow to the head of the pipe

toilet installation due to the rapid transition from expelled air to water during operation, Swaffield, Ballanco and McDougall (2002), analysed via the Method of Characteristics modelling techniques described herein.

The addition of an outwards relief valve may be shown to reduce the pressure levels experienced on gas compression, Figure 5.3. The introduction of such relief valves introduces questions as to the response rate of such devices and most importantly their closure integrity. It is recognised that in many application the use of outwards relief valves is inappropriate due to potential contamination effects, however they have been used effectively to alleviate trapped air generated surges, as reported by Ballanco (1998) and subsequently confirmed by pressure transient simulation, Swaffield, Ballanco and McDougall (2002). Figure 5.4 illustrates the circumstances generating the surge and the predicted effect of the outwards relief valve installation.

A failure to properly vent trapped air from a building internal water supply system in a Brooklyn apartment building resulted in an initial flow of water and trapped air through a w.c. as the terminal fitting on a pressurised supply. In such a US application the normal water closet cistern is replaced by a mains flushing valve. The trapped air, passing as a major 'slug' through the flush valve and then the w.c. rim, encountered little resistance thus allowing the airflow rate to rise and hence accelerating the following water column.

The resistance of the flush valve, and the rim, may be recognised as dependent upon both the square of flow velocity and the flow density. The sudden, or at best rapidly varying, flow density change from 1.3 kg/m³ to 1000 kg/m³ as the flow moves from air through a water–air mixture to water only, results in an approximate 800-fold increase in combined valve and rim resistance and hence is the equivalent of a rapid reduction in flow velocity brought about by the resistance offered by a valve increasing during closure, effectively similar in action to the rapid closure of a valve at the w.c. location. The resulting 'waterhammer' surge was sufficient to

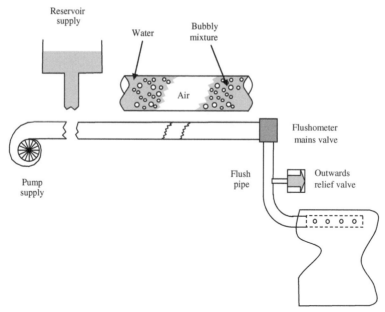

Figure 5.4 Pressure surge event fractures the w.c. ceramic as the driven trapped airflow is replaced by water flow – rim loss coefficient effectively rises by a factor of 800 and water flow velocity is instantaneously reduced causing destructive surge pressure levels

fracture the ceramic w.c. and cause major injury to a child who had just operated the flush valve. Figures 5.5 and 5.6 illustrate the experimental modelling of the event undertaken by Ballanco (1998) at Stevens Institute, Hoboken, New Jersey, subsequently supported by the surge simulations included in Swaffield, Ballanco and McDougall (2002). The addition of an outwards relief valve between the w.c. rim and the flush valve provided the required surge protection, again confirmed by both laboratory testing and surge simulation, Ballanco (1998) and Swaffield et al. (2002), Figure 5.7.

Ceramic portion of w.c. blown across reinforced transparent test enclosure by pressure surge

Figure 5.5 W.c. fractures due to the surge generated as the water flow is re-established through the rim (courtesy of J.B. Engineering)

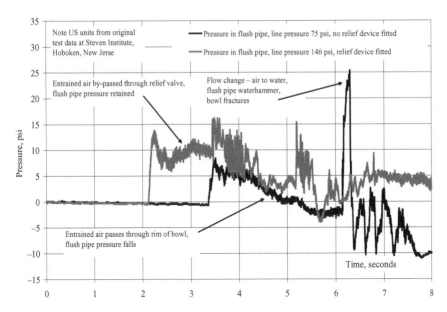

Figure 5.6 Fracturing w.c. during flush valve operation; the surge pressure generated as the accelerated water column is retarded is sufficient to 'explode' the w.c. (courtesy of J.B. Engineering)

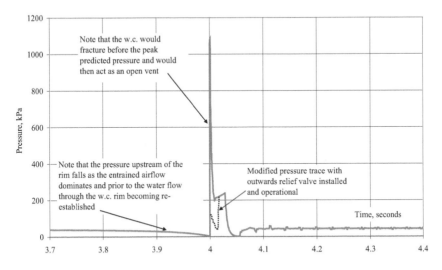

Figure 5.7 The surge pressure generated as the accelerated water column is retarded is reduced by the addition of an outwards relief valve upstream of the rim

Therefore the destructive nature of transient propagation must be recognised and strategies and equipment devised and implemented to control and suppress the transients. In the treatment of traditional pressure surge such strategies include the installation of air chambers to control both positive and negative surges, the provision of inwards relief valves to control negative pressures and outwards relief to control positive surges, although environmental issues may limit these latter applications. Pressure surge may be destructive and a full understanding of the system pressure response to both controlled or inadvertent operating point change is essential. However it would be incorrect to assume that such issues are restricted to large-scale networks delivering the widest range of fluid flows, from undersea oil pipelines to municipal sewage pumping mains and in-flight refuelling to water supply networks. The mechanisms of pressure transient propagation and the resulting system pressure response for building services installed systems obey the identical laws of fluid mechanics, whether the concern is for the pressures generated in dry risers due to insufficient venting, the surges that would accompany the sudden flow transition as expelled air driven out of the network by w.c. flushing gives way to the following water flow, or the low amplitude air pressure transients capable of depleting the standard 50 mm trap seal and facilitating cross-contamination of the habitable space.

This chapter will therefore consider control and suppression strategies and introduce the necessary equipment to limit system pressure response.

5.2   Control of transients

The objective of any transient analysis is the prevention, control and suppression of the transient. Therefore a number of fundamental 'rules' for surge protection arise from the fundamental work on transients over the past century and the enhanced understanding of transient propagation and the simulation capability provided by the Method of Characteristics.

The first rule, derived directly from Joukowsky's analysis, is that surge protection depends on reducing the rate of change of the flow conditions, Equation 2.2. In the case of flow reduction leading to positive surge pressures the techniques available would include reducing the speed of valve closure, providing a controlled valve closure so that the early stages are fast but the final close off stage is slow, or providing an alternative route for the flow, perhaps via an outwards relief valve, and/or a mechanism for reducing its deceleration.

In the case of the negative surge waves propagating downstream of a closing valve, or tripped pump, the techniques available would again include slowing the valve motion, slowing the pump shut down by means of increased inertia or providing a source of continuation flow from an air chamber or inwards relief valve, or via a valve or pump bypass, Figure 5.8.

Thus the first rule is to provide a mechanism to 'slow down' the rate of flow velocity change. It will be seen later that this rule underpins the most common 'active air pressure transient control strategies', whether through the provision of increased local inflow, as in the case of an air admittance valve, or providing

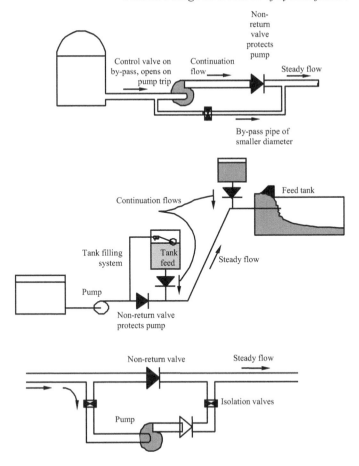

Figure 5.8 The provision of a continuing flow to minimise the rate of change of the system boundary conditions is an important surge protection design option; the cases illustrated apply to pump failure or shut down and act to limit column separation effects

a diversion route for the existing entrained airflow, as in the case of the variable volume containment device or PTA discussed earlier.

However once generated the transient will propagate throughout a network and therefore it follows that to protect any section of conduit, or any branch termination from surge damage, the control device must be positioned between the source of the transient and the site to be protected. This becomes the second fundamental precept for surge control and is one that will be returned to later.

In building drainage and vent systems, surge control has traditionally been provided by 'passive in-place' parallel venting, however it is clear that while these solutions may obey the first of the control rules outlined, they cannot obey the second as any relief from the extremities of the network can only arrive after the transient has already passed every 'at risk' appliance trap seal.

In addition, reference to the transmission and reflection of transients at junctions of system conduits introduces limiting values for the effectiveness of relief devices mounted as terminations of branch connections. It has already been shown that for a three-way junction of identical conduits the reflection coefficient is 1/3 and the transmission is 2/3 of the incoming transient. Hence a relief device on one branch only affects 2/3 of the transient. However, as demonstrated in Chapter 2, if the branch connection featured a drastically reduced wave speed then the calculation is transformed with a much greater potential for surge reduction.

While historically, pressure transient analysis has been thought of as involving high pressures, or close to vapour pressure events, that lead to conduit failure or implosion generating substantial pressure levels and catastrophic pipeline failures leading to the discharge of possibly polluting or hazardous fluids, this is not the case for all applications of the available analysis techniques. System failure is not dependent upon the absolute pressure levels attained, rather it depends upon the relationship between the transient and the operating spectrum of the system.

5.3   Applications in building utility systems

Pressure surge applications within building utility systems have already been referred to, most notably in the areas of fire-fighting dry risers and sprinkler systems where trapped air exacerbates the surge. In addition the issue of fractured water closets was investigated by Ballanco (1998). In each of these cases the pressures generated indicate that the applications belong within the generally accepted definition of waterhammer events.

Therefore the transient propagation theory presented applies independently of the magnitude of the transient. The operation of building drainage and vent systems displays identical air pressure transient propagation properties to those already described. The equations derived from the seminal work of Joukowsky, and later international researchers, apply equally to transient propagation in circumstances where a transient with an amplitude of 100 mm water gauge may trigger system failure.

The operation of building drainage networks within large complex buildings is primarily intended to ensure the efficient removal of both fluid and waste matter from the installed appliances. In addition, the venting provided is intended to ensure that there is no ingress of contaminated air or sewer gases into the building habitable space through the connected appliances. Changes in system water flows result in changes in entrained airflow, these changes being propagated throughout the drainage and vent system by means of low amplitude air pressure transients. However, while these transients are of low amplitude – 100 mm water gauge would be a severe transient – they are capable of destroying the system protection against the ingress of contaminated air provided by the appliance water trap seals. Trap seal loss may occur as a result of either negative transient propagation, typically associated with increases in annular water downflow in the vertical stacks, or positive air pressure transients, associated with interruptions to the entrained airflow through the network as a result of water flow surcharge or other sewer

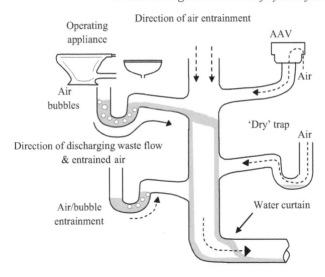

Figure 5.9 Entrained airflow path through a drainage and vent system under normal operating conditions

generated positive pressure transients. It is also recognised that trap seal loss may be caused by appliance self-siphonage, or the drying out of traps due to lack of use, high temperatures within the space served or poor maintenance.

The venting system necessary within the overall design of a complex single or multi-storey building drainage network establishes paths for an airflow from the habitable space to either the external atmosphere or the sewer system. Figure 5.9 shows the normal mode of operation of a single stack system and includes examples of a 'dry trap' that has lost sufficient water to allow air (or gaseous contaminant) movement into the network. Air is also naturally entrained by appliance operation. This assumes that the prevailing pressure regime within the system is principally negative, thereby providing a 'safe' exit route for the air through the sewer connection.

The loading of the drainage network can generate positive pressure transients that also propagate throughout the system. The cyclic nature of the water curtain at the base of the vertical stack results in interruptions to the entrained airflow and these may be sufficient to establish a contamination path through a depleted or compromised trap. Figure 5.10 illustrates this effect and demonstrates that the resultant positive pressure wave generated when the entrained air path is closed introduces the potential for air to exit into habitable spaces via a depleted trap that allows free air movement or bubble ejection.

5.4   Applications within building drainage and vent systems

Three sources of transient propagation may therefore be identified within building drainage and vent systems. Increases in applied waterflow generate increased

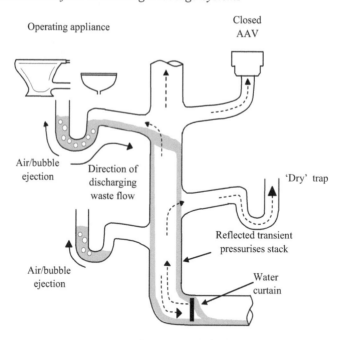

Figure 5.10 Route for cross-contamination

entrained airflows and hence propagate negative transients. System surcharge interrupts the entrained airflow and hence propagates positive transients. Finally, applied external air pressures generate transients that propagate throughout the network, for example wind shear over the building roof termination can set up sinusoidal pressure oscillations while air pressure surges within the sewer network, caused by surcharge downstream or pump selection, may generate transients, both positive and negative, that enter the system.

Figure 5.11 illustrates the simulation of transient propagation following an appliance discharge into a system where the roof termination has become inadvertently blocked. Negative surges within the network lead to the depletion of a branch trap seal and hence system failure. The introduction of an inwards relief valve, known as an air admittance valve or AAV, Figure 5.12, is seen to remedy the failure. The AAV opens as soon as the local pressure falls below a preset minimum, closing again to prevent leakage as the local pressure exceeds atmospheric.

The inadvertent blockage of the upper vent termination was therefore the cause of the trap seal loss experienced without the protection provided by the AAV installation, an example that bears similarities to the fatal asphyxiation of two passengers on a ferry when the bilge vent was blocked and en-suite traps were lost.

Figure 5.13 illustrates the operational characteristic of an AAV. Negative pressure, below a preset limit, in the stack or branch connection allows an

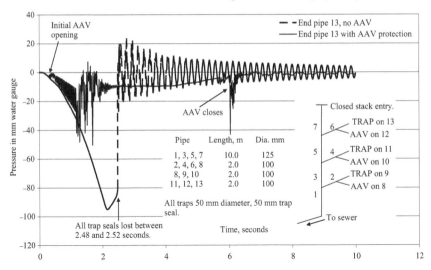

Figure 5.11 Negative transient propagation in an inadvertently blocked system alleviated by the introduction of AAV protection adjacent to each appliance trap

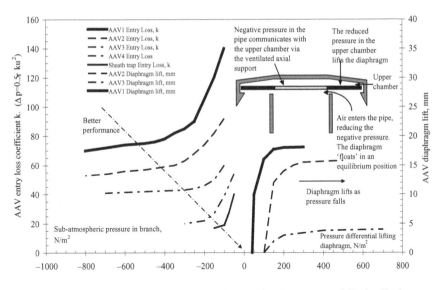

Figure 5.12 Operation of an air admittance valve. As the pipe pressure falls the diaphragm lifts and air enters the pipe. As the pressure differential increases, the diaphragm reaches its maximum displacement and the valve loss coefficient reaches a minimum.

entrained airflow to enter the network. The arrival of a positive transient at the AAV results in automatic closure, as shown.

The criteria for successful AAV operation may therefore be identified as a fast reaction to reductions in local pressure below the preset sub-atmospheric

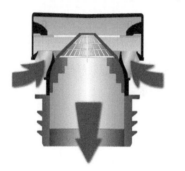

Figure 5.13 Operation of an air admittance valve (courtesy of Studor Ltd)

activation level, a low loss coefficient while open to maximise the relief airflow, and a reliable seal when the air pressure recovers within the network to prevent leakage into surrounding habitable space. To these attributes must be added a minimum maintenance requirement. While fast reaction is a requirement, it is also necessary for the moving diaphragm of the unit to be sufficiently robust to prevent flutter or sequential high frequency opening and closing. It will be seen that the opening and closing of the AAV propagate identifiable transients into the network, Figure 5.11.

Figure 5.12 also includes the pressure differential versus throughflow for a sheath trap, of the type represented by the Hepworths HepV0 unit illustrated in Figures 5.14 and 5.15. Here a sheath acts as both a trap seal to prevent ingress of air from the drainage and sewer network while under the action of either positive water pressure, applied from an appliance above the sheath, or negative air pressure in the drainage network, the sheath opens to allow the passage of waste water or entrained air. Thus the sheath device acts as both a trap seal and an AAV. Application of these waterless traps was considered as an immediate answer to the Amoy Gardens SARS cross-contamination event, see Chapter 6.

The criteria for a surge attenuator to handle positive air pressure transient propagation may also be defined by reference to the essential surge suppression and control concepts discussed previously. The most important criterion has to do with the prevention of leakage into habitable space. This dictates that whatever attenuator is designed must not allow sewer gas or any other contaminated airflow to exit the drainage network. Thus a simple outwards relief valve is excluded unless it discharges outside the building in a sufficiently remote location that re-entry of displaced gas is not possible. This therefore implies a containment vessel, Figure 5.16.

The percentage of an incoming transient that may be diverted into a branch mounted relief device depends upon the junction area ratios, see Chapter 2. If the branch is identical to the main conduit then the transmitted transient is 2/3 of the incoming wave. However if the wave speed in the containment vessel is reduced then the transmitted transient is also reduced as the transmission coefficient for three geometrically identical conduits with one having a much reduced wave speed becomes 2 / (2 + (a number >1)). Thus the choice of containment vessel

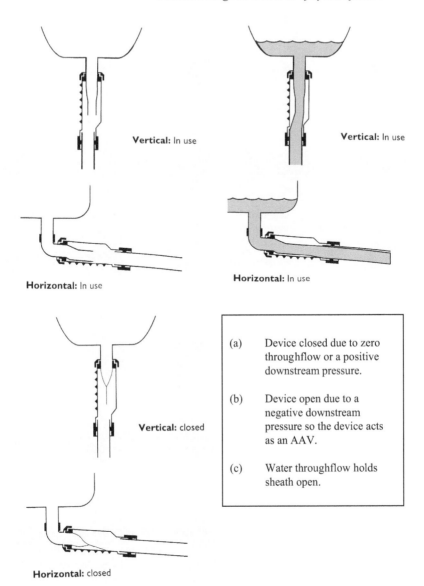

Vertical: In use

Vertical: In use

Horizontal: In use

Horizontal: In use

Vertical: closed

(a)   Device closed due to zero throughflow or a positive downstream pressure.

(b)   Device open due to a negative downstream pressure so the device acts as an AAV.

(c)   Water throughflow holds sheath open.

Horizontal: closed

Figure 5.14 Operation of the Hepworths waterless trap, illustrating the internal mounting of the sheath and its response to both water throughflow and the air pressure regime within the drainage network (courtesy of Wavin UK)

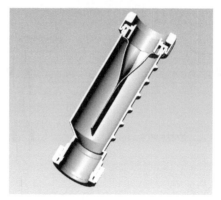

Figure 5.15 Internal construction of the Hepworths waterless trap, illustrating the internal mounting of the sheath (courtesy of Wavin UK)

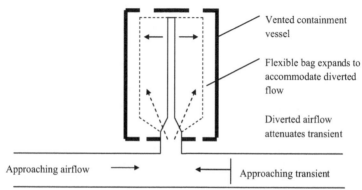

Figure 5.16 Schematic of a possible surge attenuator design to reduce the air pressure transient generated by drain surcharge

material becomes an issue as it will be necessary for this to be as flexible as possible, consistent with durability.

The earlier discussion of the effect of trapped air also has a bearing on the possible design of an attenuator. Put simply, the inflow of a diverted airflow into a zone containing gas at line pressure will immediately cause the zone pressure to rise and will therefore limit its attenuating effect. Discharging diverted airflow into an expanding volume will however not raise the zone pressure and will allow the continuation of a diversionary flow route with no diminution of efficiency until the bag is fully extended.

Again, the placement of any attenuator will determine its overall impact on transient propagation. It is imperative that it be placed, in the drainage and vent system application, between the source of the transient and the appliance trap seal to be protected.

Thus a series of criteria may be developed to define the design of a positive transient pressure attenuator. These include ease of inclusion in a drainage

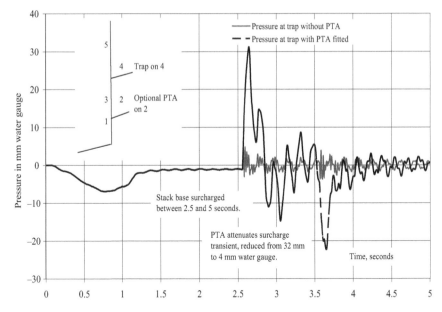

Figure 5.17 Comparative transient pressures predicted at the end of branch 4 with and without a positive transient attenuator

network, low to minimal maintenance, low inherent wave speed to maximise attenuation and minimal initial volume to prevent unit pressure increasing once a diverted airflow is established.

The simplest attenuator design to satisfy these criteria would be a zero inertia flexible bag having a zero initial volume prior to the arrival of a diverted airflow. The attenuator upper volume would be preset and once achieved the normal gas laws would govern the subsequent zone pressure.

Figure 5.17 illustrates the likely effect of such an attenuator on the transients propagated following a surcharge event in the drainage and vent system monitored in terms of the displacement of a system appliance trap seal. The simulation illustrates the effect of such a device and its impact on the transmitted transients. It will be seen that the pressure transient propagated is reduced and the trap seals previously lost or severely depleted are protected, Figure 5.18, ensuring the integrity of the drainage and vent system and preventing any egress of contaminated air or sewer gases into habitable space.

5.5   Development of a low amplitude positive transient attenuator

The basis for surge suppression and control within general pressure surge applications has been outlined, drawing upon the fundamental equations presented earlier. Similarly the prerequisites for successful surge attenuation have been demonstrated and shown to be independent of the system, the fluid and the scale of the transient.

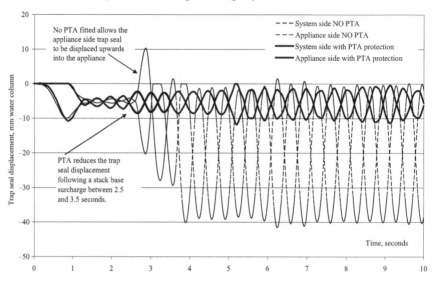

Figure 5.18 Comparison trap seal retention, branch 4 with and without a positive transient attenuator

In the particular case of the low amplitude air pressure transient propagation considered in this text the following transient drivers have been identified:

1   increases in water flow generating negative transients,
2   decreases in airflow, particularly caused by interruptions to the airpath due to stack surcharge at an offset or at the stack base, generate positive transients,
3   imposed air pressure fluctuations from outside the system, for example wind shear over the roof stack termination or sewer air pressure fluctuations arising from remote sewer surcharge or pump selection generated transients.

In each case the transients generated propagate throughout the network, being reflected and transmitted at each and every boundary, internal or terminal, and with magnitudes defined by the governing equations discussed and the principle of wave superposition demonstrated earlier.

All of these circumstances generate transients that could displace the appliance trap seals and may lead to trap seal depletion or loss and the failure of the system's ability to prevent cross-contamination of habitable space. Trap seal loss as a result of excessive negative air pressure transients may be avoided by inwards venting provision, either via a traditional dedicated vent to atmosphere or via an air admittance valve. Trap seal loss as a result of trap water displacement into the appliance as a result of the arrival of a positive pressure transient cannot be avoided by the use of an AAV as these relief valves close in response to positive pressure to avoid contaminated air ingress into habitable space, although there is evidence to support the theoretical expectation of a negative transient propagated as the AAV diaphragm closes, i.e. Figure 5.11 at 6 seconds.

The fundamental principles of transient alleviation however suggest a suitable relief device. It is necessary to reduce the rate of change of the entrained airflow and this can be accomplished by diverting the flow into a storage volume. However to develop a suitable attenuator it is first necessary to be able to simulate the propagation of low amplitude air pressure transients within a drainage and vent system and then evaluate, both empirically and by analysis, the performance of a range of possible attenuator designs.

The most effective way to control any transient propagation is to 'slow down' the rate of change of fluid velocity. For an instantaneous airflow stoppage the pressure rise associated with a reduction in airflow velocity of 1 m/s, from Equation 2.2, is approximately 40 mm water gauge. Thus if the entrained airflow potentially brought to rest by a stack surcharge can be 'diverted' and its velocity reduced 'slowly', in terms of the system pipe period, then the positive air pressure transient generated will be reduced. The challenge therefore was to design a device that would achieve this control of airflow deceleration without being cumbersome or introducing any other transient air pressure problems.

An essential point in understanding pressure transient control is to realise that the control must be applied as close as possible to the source of the transient. In terms of a building drainage system any control device must be between the source of the transient and the first appliance trap seal encountered by the transient as it propagates through the network.

Surcharges at the base of the vertical stack generate positive air pressure waves that travel up the stack and displace every trap seal between the source of the transient and the closest open terminal boundary. Poor system design that includes offsets may also be subject to local surcharging. In this case however it is also necessary to consider the transients propagated downstream, i.e. down the stack below the surcharge location. In this case the airflow is again brought to rest but the transient will be negative providing a high probability that lower levels will suffer trap seal depletion due to induced siphonage. While any distributed AAV installation will limit the possible seal depletions, the introduction of a control device below the offset that would allow a 'make up' airflow to be supplied to the stack would reduce the rate of change of airflow and would limit the transient at its source.

Thus it is possible to develop the following principles that may lead to an efficient positive pressure transient propagation control device for building drainage systems:

1   The device must allow a continuation of airflow for a period long enough to effectively reduce the rate of change of the airflow. This may be seen to apply to both positive and negative pressure transients generated as a result of stack surcharge.
2   The device must be suitable for mounting close to the likely source of the transient and must be able to survive in a 'fit and forget' maintenance regime.
3   The device must be practical in terms of scale.

4   Modelling of air pressure transient propagation within building drainage networks must be capable of providing information on the appropriate size of the device to meet any likely system loading or stack dimension design case.

As suggested above, the most efficient way to control air pressure transients is to maintain flow and reduce the rate of change of flow velocity. Both these objectives may be accomplished by diverting airflow into a control volume whose internal pressure would be allowed to increase, thus reducing the airflow deceleration from that appropriate following a sudden surcharge. Clearly such devices would operate in both the positive and negative transient cases; in the latter, flow is maintained as air is allowed to flow out of the control volume as the pressure in the connected stack or branch drops. Figure 5.19 illustrates various devices that would meet these requirements, namely a fixed volume device, an attenuating trap device and a variable volume containment device. In all cases airflow will enter the device as a result of a positive pressure transient arriving at the stack to device connection. Any initial trapped gas in the device is compressed, thus both absorbing the continuing airflow and raising the volume pressure, thereby providing flow deceleration. In the case of the variable volume containment device, if this is initially at a negative line pressure, the air inflow first opens and expands the 'bag' before any subsequent pressurisation.

In order to assess these alternate designs the existing Method of Characteristics simulation, AIRNET, Swaffield (2006), was used to simulate the effect of each of the possible devices. Each device is represented by a boundary equation, in the case of the fixed volume device the gas laws provide a boundary expression provided that the initial gas volume within the control device is known at atmospheric pressure and that the inflow to the device can be approximated from the known and to be calculated air velocity at entry to the device:

$$Vol_{t+\Delta t} = Vol_t + \Delta t\, 0.5\, A_{entry\ junction} (V_{t+\Delta t} + V_t) \tag{5.1}$$

$$P_{t+\Delta t} = Vol_{t+\Delta t} P_{atm} / Vol_0 \tag{5.2}$$

These equations are solved iteratively with the available C^+ characteristic at the entry to the device.

In the case of the attenuating trap device the gas laws relating to the volume of gas trapped above the water line are solved with the manometer equations provided the water depth in both the inner and outer zones, the initial gas volume at atmospheric pressure and the diameters of the inner and outer cylinders are known, along with the C^+ and the wave speed/pressure expression, Equation 3.28:

$$Vol_{t+\Delta t} = Vol_t + 0.5\, A_{outer} (h_{outer,\, t+\Delta t} + h_{outer,\, t}) \tag{5.3}$$

$$h_{outer} = h_{inner} (A_{inner} / A_{outer}) \tag{5.4}$$

$$P_{vol} + \rho g\, 0.5\, (h_{outer,\, t+\Delta t} + h_{outer,\, t}) = P_{t+\Delta t} + \rho g\, 0.5\, (h_{inner,\, t+\Delta t} + h_{inner,\, t}) \tag{5.5}$$

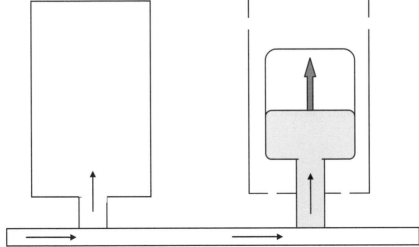

Fixed volume device – a sealed 50 litre container.

Variable volume containment device – a sealed 50 litre container, containing a flexible bag.

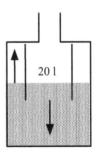

Type 1

Type 2

Attenuating trap devices type 1 and 2.

Two sizes were used, each with an internal sleeve containing a sealing water depth.

Figure 5.19 A range of possible positive air pressure transient control devices

$$P_{vol} = Vol_{t+\Delta t} P_{atm} / Vol_0 \tag{5.6}$$

For the variable volume device the boundary equation is formed by assuming the bag remains at atmospheric pressure until the accumulated air inflow exceeds its full capacity. The gas laws then apply as the device effectively becomes a fixed volume device. If the bag is completely deflated than the appropriate boundary equation represents zero outflow from the bag, effectively a one-way dead end and the opposite to an AAV.

$$P_{t+\Delta t} < 0, \qquad u_{t+\Delta t} = 0 \tag{5.7}$$

$$P_{t+\Delta t} > 0, \qquad p_{bag} = atm \tag{5.8}$$

$$P_{t+\Delta t} > 0, \qquad Vol_{bag,t+\Delta t} = Vol_{bag,\ t} + \Delta t\ 0.5\ A_{entry\ junction} (V_{t+\Delta t} + V_t) \tag{5.9}$$

$$Vol_{bag,t+\Delta t} > Vol_{full\ bag} \qquad P_{bag,\ t+\Delta t} = Vol_{bag,\ t+\Delta t} P_{atm} / Vol_{full\ bag} \tag{5.10}$$

Simulations based on these boundary conditions were then compared with a laboratory evaluation using a simulated 63 m high vertical stack, 75 mm in diameter, including connections to five simulated appliances each fitted with a 110 mm water trap seal, Figure 5.20, to avoid trap seal loss under test conditions. Airflow was generated through the stack by operation of a fan at the simulated base of the stack. Initial airflow measurements were taken in a suitably long 50 mm diameter pipe upstream of the fan. Positive air pressure transients were generated by an in-line valve closure adjacent and upstream of the fan. Air pressure measurements were recorded using a number of pressure transducers connected to a high scan rate data logging system.

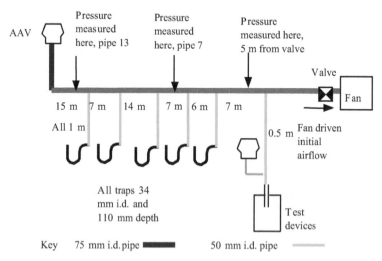

Figure 5.20 Laboratory determination of the efficiency of a range of positive air pressure transient attenuators

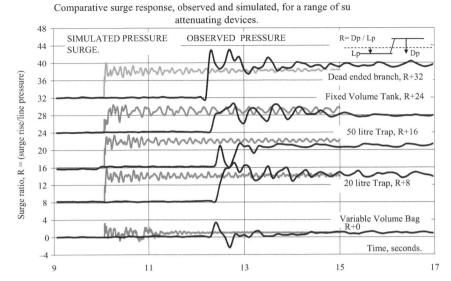

Figure 5.21 Comparative surge response, observed and simulated, for a range of surge attenuating devices (note an arbitrary increment has been added, $R = R + 8$, to separate the curves vertically)

Figure 5.21 illustrates the pressure transient response to a rapid cessation of airflow, representative of a stack surcharge downstream of the attenuator position, together with the predictions provided by the transient simulation. The surge comparisons are presented in terms of a surge ratio defined as the ratio of surge pressure change to line pressure:

$$R = Dp_{surge} / L_p \qquad (5.11)$$

(Note that an arbitrary increment has been added, $R = R + 8$, to separate the curves vertically in Figure 5.21 and the experimental and simulated results have also been separated on the time axis.)

Overall there is an acceptable level of agreement between the experimental and simulated results, providing a degree of confidence in the Method of Characteristics simulation to be used later to determine the likely efficiency of the attenuator design chosen. The measured and simulated pressure magnitudes agree well, however the measured data demonstrates more rapid damping. This is common in Method of Characteristics models of transient attenuation and is due to a range of factors, including movement of the test pipework and the actual dependence of frictional damping on the unsteady airflows present during transient propagation – contrasted to the quasi-steady flow frictional assumptions inherent in the solution. The overall agreement is considered satisfactory.

The effectiveness of the fixed volume tank device, Figure 5.19, is illustrated in Figure 5.21 – the pressure transients within the stack are reduced by some 25 per cent relative to the initial stack-only simulation. This device therefore has some potential, however the large volume necessary may be detrimental. It should be noted that the tank pressurisation commences as soon as the transient arrives. The device would clearly conform to a 'fit and forget' maintenance regime. Care would have to be exercised to ensure that water flow from the stack could not enter the tank and reduce its operational volume.

The attenuating trap devices have little apparent effect due to the mass of trap seal water that is required to oscillate as a consequence of the transient pressures in the stack. It was concluded that these devices were not suitable for further investigation.

Figure 5.21 also illustrates the operation of the variable volume bag positive transient attenuator – this is clearly a satisfactory solution as the pressure transients are reduced to zero. Effectively the flexible bag allows air to leave the stack under positive pressure conditions, this air re-entering the stack once a negative pressure regime is re-established. As long as the bag is not fully inflated the transient is discharged to a constant pressure zone at atmospheric pressure. If the bag becomes fully inflated then it pressurises in the same way as the fixed volume attenuator, losing efficiency in this mode of operation. The design objective would be not to fully inflate the bag and to ensure that it is installed in a zone with an initial line pressure below atmosphere. If the bag is completely evacuated then the attenuator to stack junction becomes a dead end.

Under normal stack operation the prevailing pressure would be below atmospheric so that the flexible bag would be collapsed, its most efficient condition to attenuate any incoming positive transients.

In each case the simulation predictions confirm the experimental study. Figure 5.22 allows the sequence of events with and without a positive air pressure attenuator in place to be considered. The test equipment is as detailed in Figure 5.20.

Closure of the valve upstream of the fan generates a positive transient, 110 mm water gauge as recorded passing the transducer. Subsequent data collection is dependent upon the reflection of this wave at the AAV termination and the closed test valve at the base of the test stack and the relevant system pipe periods. The distance from the pressure transducer to the AAV termination is 51 m and from the transducer position to the test valve 5 m. Hence the relevant pipe periods at 340 m/s wave speed are $102/340 = 0.3$ seconds to return from the AAV boundary and $10/340 = 0.03$ seconds to return from the closed test valve. On valve closure the pressure rises at the transducer in a manner typical of a ball valve closure – initially gradually and finally rapidly. The arrival of the initial transient generated by the valve closure may be seen to be at around 15.3 seconds and therefore the first reflection from the AAV boundary should be visible at around 15.6 seconds.

The form of the subsequent pressure trace, with no attenuator fitted, cannot be explained by simply assuming the AAV closes and forms a closed end as the pressure trace during the subsequent pipe periods, 15.5 to 16.0 seconds, demonstrates the arrival at the transducer of both positive and negative transients.

It is helpful to view the initial wave as being made up of three serial sections: an initial 10 mm water gauge transient that increases the pressure to the AAV closure

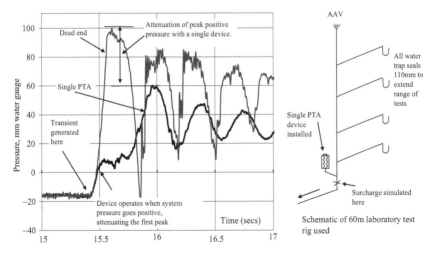

Figure 5.22 Air pressure transient propagation comparison with and without a 4 litre positive transient attenuator (PTA) fitted

threshold and initiates AAV closure, a second section that covers the closure period of the AAV and a final section that raises the pressure further at the closed AAV.

As the leading edge of the initial wavefront reaches the AAV it is still open and admitting airflow. It remains open until the pressure at the diaphragm exceeds its opening pressure. This initial positive wavefront is therefore subject to a negative reflection coefficient at the open AAV having a value less than –1 as the AAV represents a concentrated loss. A small negative pressure wave is thus propagated downstream from the closing AAV and is recorded between 15.6 and 15.7 seconds at the transducer.

As the pressure at the AAV reaches its closure threshold the diaphragm shuts, reducing the air inflow to zero and generating a negative pressure transient that propagates down the stack, determined to be –110 mm water gauge by the Joukowsky relationship and measurements of initial airflow velocity. The leading edge of this wave arrives at the transducer at 15.7 seconds, corroborated by the marked downturn in the trace. The duration of this wavefront depends on the closing time of the AAV. The initial positive transient generated by the original valve closure continues to arrive at the AAV and is reflected as a positive wave as the AAV is closed. This +1 reflection propagates a wave that follows the Joukowsky pressure drop arriving from 15.8 seconds. Due to the principle of superposition of waves the net decrease in pressure due to the transients propagated downstream during AAV closure has to be determined from a summation of these three separate waves, the actual values depending upon the speed of AAV closure and the rise time of the initial transient.

Waves arriving at the transducer from the AAV pass and are re-reflected, with *a* +1 reflection coefficient, by the closed valve immediately upstream of the fan. The leading edge of each successive wave returns to the transducer 0.03 seconds later, indicating that a summation of six waves is necessary at the transducer to predict the form of the likely pressure trace – indicating the requirement for a computerised simulation for real networks. In summary the transducer pressure falls as the AAV closes and then rises again due to the positive wave reflection following AAV closure, both reinforced by the +1 reflection imposed by the closed test valve. Figure 5.22 corroborates this model of the transient propagation and reflection.

Similar reasoning explains the trace with an attenuator included. The transient following initial valve closure opens the entrance to the flexible bag. A small positive pressure is necessary to overcome the losses in the bag entry and to open the bag itself – note that as the stack will have been subjected to a negative line pressure the bag will have been wholly evacuated and there will be a degree of stiction in separating the material to form a containment volume. The pressure recorded will therefore represent this backpressure until either the bag pressurises or reflected transients arrive from the upper stack AAV termination from 15.6 seconds onwards. The pressure wave transmitted towards the AAV, a rise of 30 mm water gauge above line pressure, Figure 5.22, will increase the pressure at the AAV and close it. However the resulting negative wave generated as before by the cessation of the air inflow will immediately reduce the pressure to the AAV opening level. The AAV will hunt, allowing air to enter the stack. The pressure at the AAV will therefore be held below atmosphere, propagating a negative wave back towards the transducer that will record the arrival of a series of small oscillatory pressure waves. However as the AAV remains open, the continuing airflow into the system will fill and pressurise the variable volume bag, increasing the pressure at the transducer location.

The pressurisation of the flexible bag brings the flow to rest and propagates a positive pressure transient towards the AAV, resulting in its final closure and an identical series of reflections and transmissions to those already described for the system without an attenuator in place. This is confirmed by the period of the subsequent pressure oscillations that match those with no attenuator. Replacing the sudden valve closure with the pressurisation of the flexible bag accounts for the rather more gradual and 'rounded' pressure oscillations in the second case.

Figure 5.22 indicates residual trapped line pressure at the end of the event. In both cases the end result is a system closed at both upper and lower terminations. The residual pressure is therefore a function of the airflow that entered the system between initial test valve closure and the final closure of the AAV, reduced by the air that passed through the closing test valve during its closure.

The assessment of the mode of operation of the attenuator therefore provided a means of assessing the efficiency of any containment volume design. The relative magnitude of the first peak, defined in Figure 5.22, with and without an attenuator clearly provides a measure of efficiency. Figure 5.23 illustrates first peak data for a wide range of bag materials. It was initially felt that material choice would be a major factor. This was not supported by prototype testing. The prime considerations were that the bag opened quickly and provided an expansion

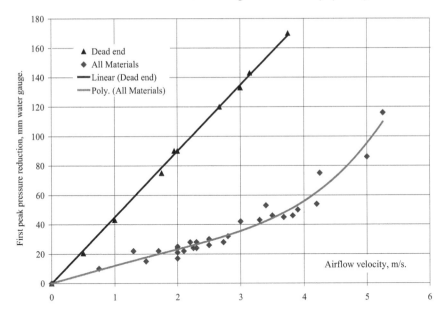

Figure 5.23 Transmitted reduced first peak for all volumes and materials compared to system with no device

volume capable of resetting itself to its collapsed state following the end of the transient event.

Figure 5.24 illustrates the effect of series mounting of the attenuator, namely direct coupling of one bag on another yielding a larger total containment volume as illustrated in Figure 5.25, a mounting arrangement incidentally inspired by Eagle (1951). While the increase in attenuation is not a linear relationship due to the convoluted air path through the device reducing its efficiency, there is sufficient evidence to support this means of providing additional volume. The data is also presented as the percentage attenuation delivered in each case as the initial airflow velocity rises, falling into three distinct zones.

At low airflows a constant attenuation of 90 per cent is achieved, limited by the backpressure necessary to open the bag, as the bag is capable of absorbing all the additional air inflow without pressurising, representing the best that could be achieved. The extent of this zone depends on bag volume and airflow velocity – up to 0.8 m/s for a single 4 litre bag rising to 2.2 m/s for two 4 litre bags in series.

Above 2.5 to 3 m/s both single and series bag configurations again deliver roughly constant attenuation irrespective of airflow. In both cases the bag volumes have become pressurised and the constant attenuation implies that, as the Joukowsky pressure rises linearly with destroyed airflow velocity, so does the maximum pressure peak. The single 4 litre bag provided an attenuation of 38 per cent and the series bag configuration 64 per cent, a ratio of 1.68, less than a doubling based on the increased volume.

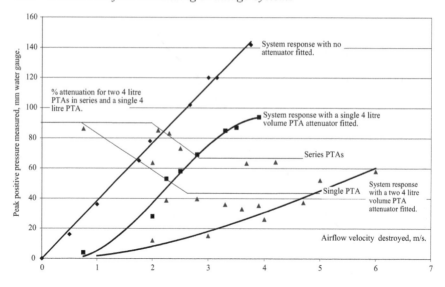

Figure 5.24 Comparison of peak positive pressures measured in the system with a single 4 litre attenuator, two 4 litre attenuators in series and a system with no attenuator fitted

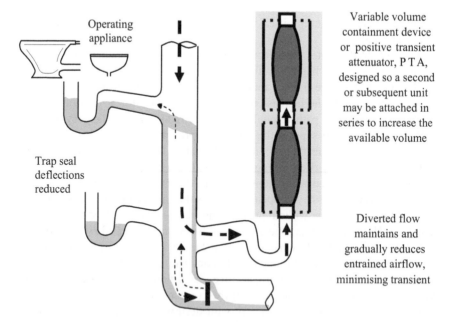

Figure 5.25 Series mounting of two PTA units to increase the containment volume

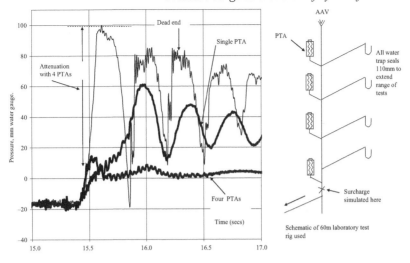

Figure 5.26 Air pressure traces following valve closure with multiple positive transient attenuators (PTA) distributed up the system vertical stack

Between these first and third zones lies an expected transition zone representing a set of flow conditions where the bag is filled to an increasing percentage of its maximum volume without pressurisation.

Figure 5.26 illustrates an alternative control choice, namely the distribution of attenuators up the vertical stack. The quadruple attenuator installation effectively destroys the pressure surge following valve closure. The total 16 litre volume clearly exceeds the air volume entering the system during and following valve closure. The airflow to the network naturally attenuates and the AAV closes gently with no surge propagation worth mentioning. This example confirms the applicability of the containment volume approach as it has effectively allowed the airflow to be diverted and decelerated gradually. As shown earlier, reducing the rate of change of air velocity is the key to surge suppression.

5.6 Active and passive pressure transient control strategies

Drainage network designers already have a range of alleviation devices to protect against negative transient propagation – traditional venting together with anti-siphon traps and AAVs provide excellent solutions. No such devices exist to deal with positive transients. The variable containment volume device offers this design flexibility and its operation has been shown to be wholly explainable in terms of pressure transient propagation theory when the attenuator is represented by a range of boundary conditions dependent upon the degree of bag pressurisation.

The introduction of both AAVs and positive transient attenuators will allow a new approach to pressure transient control and suppression in building drainage and vent systems. This concept may be referred to as active venting as opposed to the traditional passive solutions. These terms link directly to the definitions of Method of Characteristics boundary conditions already discussed. It is therefore possible to assess the practicality of this mode of pressure transient control through the simulation of system performance implementing either active or passive suppression strategies.

Active Control will be a major contributor to both reducing the complexity of venting design and providing for the first time a means of alleviating positive transients without recourse to external venting terminations. By the nature of roof terminations, the distance to be travelled by the pressure wave and any relieving reflection militates against their efficiency as such installations contravene the basic rules of surge attenuation that the attenuator must be positioned between the source of the transient and the device to be protected. By the time the relief reflection returns from a roof termination the positive transient will have adversely affected every trap seal en route. Taken together with the air admittance valve, introduced to relieve negative transients, the positive transient attenuator represents a major improvement in the range of solutions available to the drainage system designer.

The prevention of cross-contamination via depleted trap seals has been a design consideration over the past 150 years. The invention of the water seal trap in the 18th century – a 'U' bend immediately downstream of the appliance with a water depth of 50–75 mm – has remained the most effective barrier to sewer gases. Traps respond to network pressure so system failure involving cross-infection may follow the depletion of trap seals by air pressure transient propagation. Modern design, water conservation and the need to economise demands a re-evaluation of drainage design that recognises the unsteady nature of system flows and the effects of pressure transient propagation. Demands on urban living space that increase system loading due to occupation levels in excess of those envisaged at the design stage, will compromise drainage operation. Pressure transient propagation leading to system failure is associated with destructive forces in complex fluid systems. While the definition of failure is system dependent, the underlying principles of surge propagation, suppression and control remain constant. Transient propagation communicates flow demand – negative transients demand an increase in flow while positive transients reduce flow and increase pressure.

While the propagation of low amplitude air pressure transients is a natural and unavoidable consequence of appliance discharge to a building drainage system, the protection of appliance trap seals is dependent on the control and suppression designed into the system. From the late 19th century, this control and suppression depended upon fixed venting running parallel to the wet stacks. The earliest 'two-pipe' systems separated foul from general waste flows with each appliance independently vented. In the 1930s the 'one-pipe' system discharged all appliances to a common wet stack but again separately vented appliances. In the 1970s the UK introduced a 'single stack' system that dispensed with separate vents although above 30 floors a parallel vents stack cross-connected into the wet stack was introduced. All these designs featured vent stacks smaller in diameter than the wet stack and all represent 'Passive' Control and suppression as there is no interaction between the control mechanism, the fixed in place vent, and the transient. Two basic rules of surge suppression have been identified:

1   Transients may be attenuated by reducing the rate of change of flow velocity. This follows from Equation 2.2 and implies that flow should be diverted in the case of a positive transient or, in the case of a negative transient added through an adjacent inlet.
2   The surge alleviation should be positioned between the source of the transient and the equipment to be protected.

While the fixed in place vent solution provides a degree of flow diversion or addition, criteria 1 above, its efficiency in this role is limited by fundamental misunderstandings of the operating mechanism of the vent stack currently embedded in the codes.

Fixed in place vents do not meet the second criteria in any way. The source of any relief to offset the pressure regime imposed on the system by the passage of the transient is the reflection of the transient at the upper open termination of the vent system. Thus the potentially trap seal depleting transient has already passed all the traps to be protected before any relieving reflection can be generated by the open termination.

The pressure transient transmission and reflection coefficients at junctions may be determined from the following expressions, Swaffield and Boldy (1993).

$$C_{Transmission} = \frac{2\dfrac{A_1}{c_1}}{\dfrac{A_1}{c_1}+\dfrac{A_2}{c_2}+\dfrac{A_3}{c_3}} = \frac{2}{1+\dfrac{A_2}{A_1}+\dfrac{A_3}{A_1}} = \frac{2}{1+\dfrac{A_{Branch}}{A_{Incoming}}+\dfrac{A_{Continuation}}{A_{Incoming}}} \quad (5.12)$$

$$C_{Reflection} = \frac{\dfrac{A_1}{c_1}-\dfrac{A_2}{c_2}-\dfrac{A_3}{c_3}}{\dfrac{A_1}{c_1}+\dfrac{A_2}{c_2}+\dfrac{A_3}{c_3}} = \frac{1-\dfrac{A_2}{A_1}-\dfrac{A_3}{A_1}}{1+\dfrac{A_2}{A_1}+\dfrac{A_3}{A_1}} = \frac{1-\dfrac{A_{Branch}}{A_{Incoming}}-\dfrac{A_{Continuation}}{A_{Incoming}}}{1+\dfrac{A_{Branch}}{A_{Incoming}}+\dfrac{A_{Continuation}}{A_{Incoming}}} \quad (5.13)$$

It will be seen from Equations 5.12 and 5.13 that the wave speed in each pipe or duct is included in the coefficient determination, however in the case of low amplitude air pressure transient propagation in building drainage and vent systems the pipework may be taken as rigid and the wave speed in air as constant, simplifying the equations.

Similarly it will be seen that the transmission and reflection coefficients depend upon the identification of the pipe carrying the incoming transient. The junction will present different coefficients for transients arriving along the branch or the continuation pipe. Thus Equations 5.12 and 5.13 have been recast in terms of the pipe carrying the incoming transient (pipe 1 in Figure 5.27), the branch (pipe 2 in

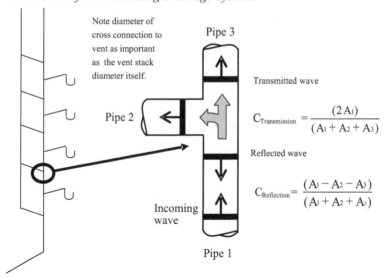

Figure 5.27 Transmission and reflection of a transient at a three-pipe junction

Figure 5.27) and the continuation pipe (pipe 3 in Figure 5.27) as this will make calculation of the coefficients easier.

The transmission coefficient at a junction of three equal-diameter pipes is 66 per cent of the incoming wave, Figure 5.28. A –33 per cent reflection of the incoming wave is also generated. If the branch vent (pipe 2 in Figure 5.27) is reduced in diameter then the transmitted wave strength increases – e.g. if the vent is half wet stack diameter then the transmitted wave is increased to 90 per cent of the incoming wave. This offers no reduction in the transient propagating up the wet stack. If the vent has a greater diameter than the wet stack then the vent

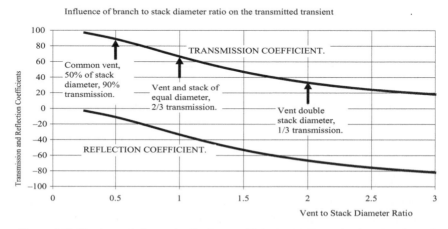

Figure 5.28 The transmission and reflection coefficients at a three-pipe junction depend upon the relative area ratios of the joining pipes; Figure 5.27 illustrates the necessary equations defining these coefficients

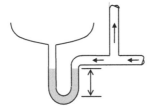

Diversion of incoming transient depends on area ratio of the vent pipe cross-sectional area to that of the trap branch

To be effective in reducing pressure applied to the trap seal the vent should be greater in cross-section than the branch

Figure 5.29 Passive vent connections applied locally to protect trap seals also require a larger vent diameter to be effective

system starts to have an influence on the transient propagated up the building, e.g. if the vent stack is double the wet stack diameter then the transmission reduces to 33 per cent. Note that the diameter of the cross-vent, Figure 5.27, is as important as the vent diameter in restricting wave attenuation.

All national plumbing codes suggest equal or smaller diameter vent stacks compared to the wet stack, hence there is a fundamental misunderstanding of the mechanism of surge protection embedded in the design codes.

It is the ratio of the pipe cross-sectional areas that determines the coefficients rather than actual pipe diameters, Figure 5.29. If the traditional passive venting of individual traps back to the vent stack is considered then it will be appreciated that a small diameter vent connected into the trap branch will have little effect.

The need to minimise external pipework and the advent of taller buildings led to the introduction of the single stack system in the 1970s. Further reductions from the mid-1980s introduced air admittance valves installed within the habitable space to allow inwards air pressure relief. Active transient control extends this approach to include both positive and negative transient suppression to provide trap seal retention and prevent cross-contamination of habitable space. Figure 5.30 illustrates an air admittance valve, AAV, and the positive transient attenuator, PTA or flexible containment volume, capable of absorbing transients until pressurised. The principle of operation of the AAV is to open whenever the local air pressure falls below a predetermined level in the local network, allowing an air inflow that

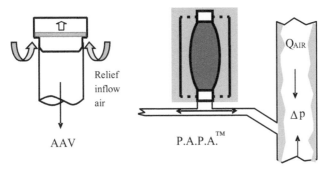

Figure 5.30 Active air pressure transient suppression devices to control both positive and negative surges

does not require the transient to travel the whole height of the building to the first roof line open termination.

The PTA allows entrained airflow to be diverted into the containment volume and reduces the rate of airflow deceleration by providing a diversion path. The pressure rise associated with the flow stoppage (Swaffield, Campbell and Gormley 2005a) is therefore reduced. Thus it may be appreciated that Active Control and suppression meet both the criteria.

5.7   Evaluation of Active and Passive Control and suppression strategies for a simulated network

Figure 5.31 illustrates a network that will allow the direct comparison of several design solutions – all stacks and branches 100 mm diameter, the trap is a 50 mm seal and the vent stack is initially 80 mm diameter. Interfloor height is 5 m. The applied water flow is a 2 litre/s flow with a 0.8 rise time from 1.0 seconds. This trial will impose a negative transient on the network and will test the ability of the Active Control AAV installation compared with various passive venting solutions with differing vent stack diameters. The base of the stack is surcharged from 3.5 to 4.0 seconds to impose a positive air pressure transient onto the network to test the ability of the Active Control PTA installation compared with various passive venting solutions with differing vent stack diameters.

Thus this single simulation includes both the possibility of induced siphonage and trap seal loss following a system surcharge dependent on system characteristics. Figures 5.32, 5.33 5.34 and 5.35 illustrate the system operational conditions for two design cases, namely an Active Control application including

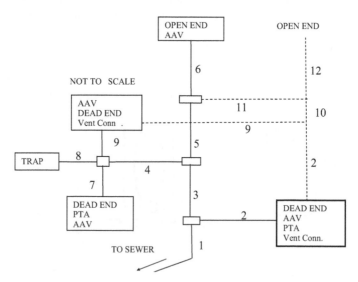

Figure 5.31 A drainage and vent system to allow the evaluation of the relative performance of an active or passive transient control and suppression strategy

distributed AAVs and a PTA at pipe 2 and a traditional scheme using an 80 mm diameter parallel vent. Trap seal water is lost as the imposition of the annular water downflow generates negative stack air pressure. Seal loss is dependent on the waterflow acceleration – 2.5 litre/s^2 is a challenging criteria. Stack base surcharge results in a positive transient propagation, however the inclusion of the PTA Active Control device prevents any additional trap seal loss. The parallel vent system does not control the positive transient and a secondary trap seal loss is experienced. Air pressure values in pipe 3 indicate that Active Control was

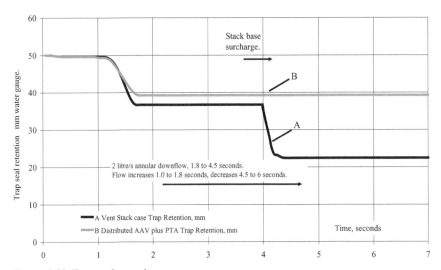

Figure 5.32 Trap seal retention

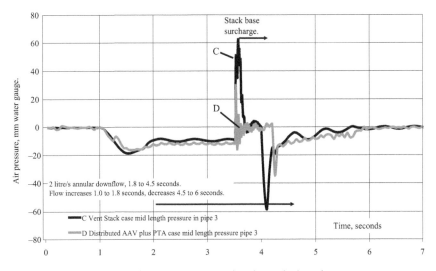

Figure 5.33 Stack base surcharge pressure transient in vertical stack

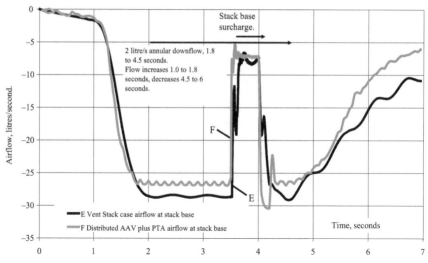

Figure 5.34 Airflow reduction during stack base surcharge event

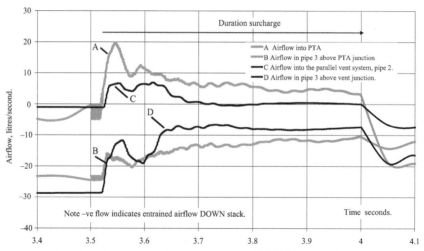

Figure 5.35 Airflow distribution with and without the P.A.P.A.™ installation

more efficient at reducing the propagated positive transient following stack base surcharge. Figure 5.36 summarises these results.

Table 5.1 compares trap seal retention and peak pressure following surcharge for all cases. Active Control results in improved trap seal retention. Introducing AAVs alone reduces the positive transients experienced as the airflow into the network is reduced and so the stack base surcharge acts on a lower entrained airflow, generating a weaker transient. Table 5.1 indicates that for a parallel vent system to have a similar performance, the vent diameter would have to be twice that of the wet stack diameter at 200 mm, a counter-intuitive result justified by the transient transmission relationship for junctions.

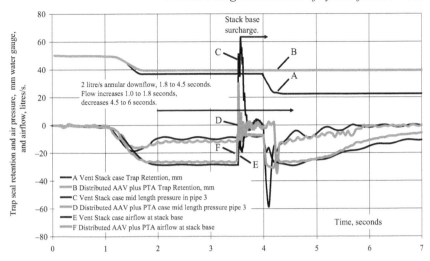

Figure 5.36 Summary of the effect of Active Control on limiting the transient pressure excursions within a building drainage and vent system

Table 5.1 Comparative system performance for various levels of Active Control and parallel vent sizing

Network description. Figure 5.31	Trap seal retention mm water gauge, at 3.5 seconds	Trap seal retention mm water gauge, at 7.0 seconds	Maximum pressure mm water gauge, mid length pipe 3
Parallel Vent Stack, 200 mm dia. with 100 mm dia. cross vents	45.68	41.25	16.85
Single Stack, Distributed AAV pipes 6, 7, 9, 3, PTA pipe 2.	39.20	39.20	28.16
Parallel Vent Stack, 200 mm dia. with 50 mm dia. cross vents	41.82	36.38	22.18
Single Stack, AAV pipes 7 and 9, PTA pipe 2	35.08	35.08	18.90
Single Stack, AAV pipes 7 and 9, PTA pipes 2 and 7	34.48	34.34	18.39
Single Stack, AAV pipe 9, PTA pipe 2	33.49	31.12	20.90
Single Stack, AAV pipe 9, PTA pipes 2 and 7	33.54	30.67	18.13
Single Stack, Distributed AAV pipes 6, 7, 9, 2	39.74	26.74	51.70
Parallel Vent Stack, 80 mm dia. with 50 mm dia. cross vents	36.70	22.44	62.40
Single Stack, AAV pipes 7 and 9 no PTA	35.08	17.44	48.41
Single Stack, PTA on pipe 2	28.07	13.32	20.83
Single Stack, AAV on pipe 9	34.00	12.82	55.94
Single Stack, no AAV, PTA or paralel vent	27.80	1.58	62.43

Reducing vent diameter increases the transmission coefficient and reduces attenuation. A 200 mm vent stack diameter reduces the transmission coefficient to 0.33 and allows greater diversion of the airflow that would have been brought to rest by the surcharge, thus conforming to the concept of surge protection already discussed – a similar but less efficient mechanism to that used by the positive transient attenuator PTA (Swaffield et al. 2005b).

The modelling capability provided by the Method of Characteristics and the application of pressure surge analysis to building drainage and vent systems presents an opportunity to re-evaluate drainage design to reduce both complexity and labour and equipment costs while providing effective protection against cross-contamination via the depletion of trap seals.

5.8   Application of Active Control to an extreme surcharge condition

Occupants of the Pak Tin estate building in Hong Kong experienced severe surcharge conditions resulting in water egress from the lower floor w.c. suites,

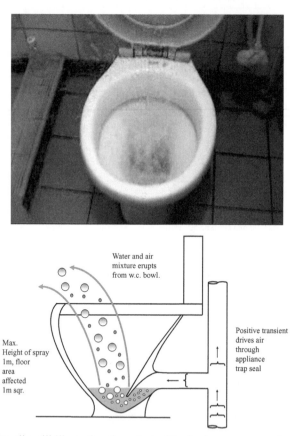

Figure 5.37 Video clip still illustrating the extent of the backpressure experienced during sewer surcharge at the base of the multiple stacks serving the Pak Tin building (courtesy of Studor Ltd) The accompanying figure illustrates the mechanism resulting in trap seal water being ejected from the w.c. as a spray with a maximum height of 1m and a floor coverage in excess of 1 m squared.

Figure 5.37. Investigations suggested that the complex multiple vertical stack connections to the sewer were operating under conditions of 100 per cent full bore flow in the sewer as well as the effects of local surcharge at the base of the building's vertical stack due to the chosen configuration of stack linking and direction change illustrated in Figures 5.38 and 5.39.

Details of the precise underground connections were unavailable, however the investigative expedient of lifting a manhole cover downstream of the vertical stack to sewer connection released trapped air pressure and confirmed the surcharge nature of the problem. In order to 'live with' the problem the occupants of the building at peak load times were known to place sandbags into the w.c.s to stop the waste entering their apartments.

The original design for the Pak Tin estate was a modified one-pipe system 100 mm stack with 50 mm vent pipes. When the reported failures had to be dealt with the initial solution was to upsize the stack from ground level to the 11th floor (the problem floors) to 150 mm diameter. This did not resolve the situation for the obvious reason that the air entrained into the system had nowhere to go once the sewer surcharged.

Prior to the invention and availability of the P.A.P.A.™ device, consultants from Studor introduced Studor maxi vent AAVs, placing them on the lower floors. This reduced the problem slightly, probably due to the negative transient generated as the AAV closed, but at peak times w.c. traps were still blown out.

The arrival of the first production P.A.P.A.™ units allowed the consultants to place a few units at the base of the stacks in the belief that the positive pressure

Figure 5.38 Multiple stack configuration at the base of the Pak Tin building (courtesy of Studor Ltd)

Figure 5.39 Detail of the multiple stack configuration entry to the below ground sewer system at the Pak Tin building (courtesy of Studor Ltd)

would be attenuated. The Pak Tin installation was the second use of the P.A.P.A.™ and the first on a problem building. It was not fully appreciated initially that at the peak loading times the below ground sewer system was at 100 per cent capacity.

Site visits identified a single point in the below ground system where up to seven stacks from the 56-floor building drain into a 'well' before entering the drainage sewer system. It was established that this was a choke point by removing an adjacent manhole cover, an action that led to the problem disappearing, however a venting solution based on this observation would not have been practical as the manhole was located at the entrance of the building.

As the severity of the back pressure was outside what could be normally expected during a positive transient event due to a short surcharge, the consultants applied a pragmatic engineering solution by increasing the number of P.A.P.A.™ units installed until the available containment volume was sufficient to attenuate the reflected airflow, Figures 5.40 and 5.41, White and Chang (2009).

Following three attempts to provide sufficient containment, a satisfactory solution was achieved with the correct amount of attenuation. This application was a very good early test of both the P.A.P.A.™ unit and the underlying positive transient attenuator theory; since 2004 there have been no reported issues due to positive pressure in these buildings indicating that retro-fitting the P.A.P.A.™ without the need to rework the below ground or above ground drainage system solved a major issue to the long-suffering occupants of the building.

Figure 5.40 P.A.P.A.™ installation at the Pak Tin estate, illustrating lower floor multiple placement (courtesy of Studor Ltd)

Figure 5.41 Detail of the multiple series P.A.P.A.™ configurations, double and triple series, used at the Pak Tin estate to accommodate the sewer surcharge pressures (courtesy of Studor Ltd)

5.9   Concluding remarks

This chapter has introduced the concepts of surge control and suppression by firstly drawing on the wider arena of pressure surge analysis and then reading across to applications within building drainage and vent system design and provision.

The concept of active as opposed to passive venting control draws upon the active and passive definitions of system boundary conditions introduced during the development of the Method of Characteristics simulation techniques. The development of Active Control devices, air admittance valves to counter negative transients and positive air pressure transient attenuators based on variable containment volume divergence of transient flows has been detailed; in the case of the P.A.P.A.™ the development of the device itself has been explained as well as its potential demonstrated through appropriate simulations and reproduced experimental and site test data.

Thus this chapter has emphasised the possibility of enhanced pressure transient control and suppression based on, and supported by, the body of pressure surge research that stretches back a century to Joukowsky in St Petersburg.

Building drainage and vent system design relies on codes that in the main have been developed from practice 'rules of thumb' or steady state experimental research, much now dated or, as demonstrated by the treatment in this chapter, predicated on a fundamental misunderstanding of the mechanisms of transient control and suppression based on passive, fixed in place, vent networks – the traditional basis of system venting. There is a need to re-evaluate the design of these networks against current criteria, including water conservation, an escalation in building complexity, increased occupation levels, enhanced concerns as to cross-contamination and ever-increasing building height. Reliance on codes is no longer sufficient. There is a need to move drainage design into the same arena as other building services system design where validated simulation techniques provide a background to allow designers and consultants to deal with applications that lie outside the specific range of cases dealt with in codes. The Method of Characteristics driven simulations presented herein, along with the Active Control design opportunities, provide a basis for this re-evaluation that rests on extensive research as well as drawing on over a century of analysis and practice in the area of pressure surge theory.

6 Application of the Method of Characteristics based simulation to a series of case studies

The application of low amplitude air pressure transient simulation to building drainage and vent system design has been shown in the earlier chapters to be feasible and to offer a range of advantages, both in allowing a better understanding of the mechanisms of system response and in aiding the development transient control and suppression strategies and equipment.

The simulation methodology may also be applied to push the design envelope and challenge previous design boundaries, making possible new approaches to real problems.

This chapter will demonstrate the application of transient simulation to three distinct conditions, in each case the presentation will outline the problem to be solved, the simulations undertaken and the design strategies that arise as solutions or future developments in approach.

The cases chosen are:

- the need to limit the use of open vent terminations in certain building designs, possibly for security considerations or simply due to other architectural imperatives;
- the need to rapidly understand the virus spread drivers in the Amoy Gardens SARS outbreak in 2003 as a precursor to a more generalised review of the security of building drainage systems in preventing cross-contamination;
- the need to build on the SARS experience, where dry traps facilitated the virus transfer, by developing a non-invasive, non-destructive remote trap seal status identification system through initial simulation and site development.

All three examples will demonstrate the applicability of the Method of Characteristics based simulations to aid real developments in building drainage design.

6.1 The sealed building – an example of enhanced air pressure transient control and suppression

The SARS virus spread, identified as reliant on dry traps in the Amoy Gardens outbreak in Hong Kong in 2003, highlighted the cross-contamination possible if dry traps occur within a building. Dependent upon the building usage, population mix and occupant home location if not a residential building, the possible commuter based infection footprint resulting from cross-contamination could be extensive. It has been demonstrated in earlier chapters that trap seal loss may occur due to the low amplitude air pressure transients propagated within the building drainage and vent system, either as a result of flow condition change or due to the effect of some inadvertent system failure, such as a blocked roof open termination. The concept of Active Control, introduced in Chapter 5, provides a possible solution to the latter case as it may not be necessary to provide open roof level vent terminations provided the designer fully understands the concepts of Active Control and diversity.

It may also be the case that architectural design constraints introduce limits on the availability of traditional open roof terminations, a current example being the flexible roof structure of the O_2 arena in London that cannot be penetrated by vent stacks.

Thus the application of Active Control for both potential security and architectural reasons is worthy of further investigation.

Air pressure transients generated within building drainage and vent systems as a natural consequence of system operation may be responsible for trap seal depletion and cross-contamination of habitable space, WHO (2003, 2004). Traditional modes of trap seal protection, based on the Victorian engineer's obsession with odour exclusion, Billington and Roberts (1982), Finer (1952) and Teale (1881), depend predominantly on passive solutions where reliance is placed on cross-connections and vertical stacks vented to atmosphere, Wise (1957) and BSI (2002). This approach, while both proven and traditional, has inherent weaknesses, including the remoteness of the vent terminations, Pink (1973b), leading to delays in the arrival of relieving reflections, and the multiplicity of open roof level stack terminations inherent within complex buildings. As shown earlier there is also the issue of misinterpretation of the action of the vent stack – traditional vent stacks are of lesser diameter than the wet stack and hence have a limited transient attenuation capability. The complexity of the vent system required also has significant cost and space implications, Wyly and Galowin (1985).

The development of air admittance valves (AAVs) over the past two decades provides the designer with a means of alleviating negative transients generated as random appliance discharges that contribute to the time dependent water flow conditions within the system. AAVs represent an Active Control solution as they respond directly to the local pressure conditions, opening as pressure falls to allow a relief air inflow and hence limiting the pressure excursions experienced by the appliance trap seal, Swaffield, Jack and Campbell (2004).

However AAVs do not address the problems of positive air pressure transient propagation within building drainage and vent systems as a result of intermittent closure of the free airpath through the network or the arrival of positive transients generated remotely within the sewer system, possibly by some surcharge event downstream – including heavy rainfall in combined sewer applications.

The development of variable volume positive transient attenuators (PTAs), Swaffield, Campbell and Gormley (2005b), which are designed to absorb airflow driven by positive air pressure transients, completes the necessary device provision to allow active air pressure transient control and suppression to be introduced into the design of building drainage and vent systems, for both 'standard' buildings and those requiring particular attention to be paid to the security implications of multiple roof level open stack terminations. The PTA consists of a variable volume bag that expands under the influence of a positive transient and therefore allows system airflows to attenuate gradually, therefore reducing the level of positive transients generated. Together with the use of AAVs, the introduction of PTA units, as demonstrated by the first such marketable device, Studor's P.A.P.A.™, allows consideration of a fully sealed building drainage and vent system.

Both generic devices have been fully described and their actions demonstrated in previous chapters through the application of Method of Characteristics based simulation.

Active air pressure transient suppression and control therefore allows for localised intervention to protect trap seals from both positive and negative pressure excursions. This has distinct advantages over the traditional passive approach. The time delay inherent in awaiting the return of a relieving reflection from a vent open to atmosphere is removed and the effect of the transient on all the other system traps passed during its propagation is avoided.

In complex building drainage networks the operation of the system appliances to discharge water to the network, and hence provide the conditions necessary for air entrainment and pressure transient propagation, is entirely random. No two systems will be identical in terms of their usage at any time. This diversity of operation implies that inter-stack venting paths will be established if the individual stacks within a complex building network are themselves interconnected. It is proposed that this diversity be utilised to provide venting and to allow serious consideration to be given to sealed drainage systems. It is likely that the interconnections required would be at the highest level, i.e. within that sector of the system that would be normally described as the dry stack region.

In order to fully implement a sealed building drainage and vent system it would be necessary for the negative transients to be alleviated by drawing air into the network from a secure space and not from the external atmosphere. This may be achieved by the use of air admittance valves local to appliances or at a predetermined location within the building, for example an accessible loft space.

Similarly, it would be necessary to attenuate positive air pressure transients by means of PTA devices mounted within the building envelope. Initially it might be considered that this would be problematic as positive pressure could build within the PTA installations and therefore negate their ability to absorb transient airflows. This may again be avoided by linking the vertical stacks in a complex building and utilising the diversity of use inherent in building drainage systems as this will ensure that PTA pressures are themselves alleviated by allowing trapped air to vent through the interconnected stacks to the sewer network.

Diversity also protects the proposed sealed system from sewer driven overpressure and positive transients. A complex building will be interconnected to the main sewer network via a number of connecting smaller bore drains. Adverse pressure conditions will be distributed and the network interconnection will continue to provide venting routes.

The propagation of air pressure transients within building drainage and vent systems belongs to a well-understood family of unsteady flow conditions defined by the St Venant equations of continuity and momentum, and solvable via a finite difference scheme utilising the Method of Characteristics technique. Air pressure transient generation and propagation within the system as a result of air entrainment by the falling annular water in the system vertical stacks and the reflection and transmission of these transients at the system boundaries, including open terminations, connections to the sewer, appliance trap seals and both AAV and PTA Active Control devices, may be simulated with proven accuracy. The simulation, Swaffield and Jack (2004), provides local air pressure, velocity and wave speed information throughout a network at time and distance intervals as short as 0.001 seconds and 300 mm. In addition, the simulation replicates local appliance trap seal oscillations and the operation of Active Control devices, thereby yielding data on network airflows and identifying system failures and consequences. While the simulation has been extensively validated, its use to independently confirm the mechanism of SARS virus spread within the Amoy Gardens outbreak in 2003 has provided further confidence in its predictions, Jack, Swaffield and Filsell (2004).

The application of the method of characteristics to the modelling of unsteady flows was first recognised in the 1960s, Lister (1960). The relationships defined by Jack (2000) allows the simulation of the traction force exerted on the entrained air. Extensive experimental data allowed the definition of a 'pseudo-friction factor' applicable in the wet stack and operable across the water annular flow / entrained air core interface to allow combined discharge flows and their effect on air entrainment to be modelled. The airflow entrained in the lower levels of the wet stack may exceed that appropriate to the annular water flows present in the upper levels. The variable friction factor allows the lower level annular flow to provide airflow entrainment and hence a rising air pressure, while in the upper levels this air is effectively drawn down past a slower moving water film that impedes its entrainment, and leads to an observed reduction in air core pressure levels.

These concepts will be demonstrated by a multi-stack network that will be subjected to both appliance discharge driven and imposed sewer transients.

6.2 Simulation of the operation of a multi-stack sealed building drainage and vent system

Figure 6.1 illustrates a four-stack network. The four stacks are linked at high level by a manifold leading to a PTA and AAV installation. Water downflows in any stack generate negative transients that deflate the PTA and open the AAV to provide an airflow into the network and out to the sewer system. Positive pressure generated by either stack surcharge or sewer transients are attenuated by the PTA and by the diversity of use that allows one stack to sewer route to act as a relief route for the other stacks.

The network illustrated has an overall height of 12 m. Pressure transients generated within the network will propagate at the acoustic velocity in air: 330 m/s. This implies pipe periods, or round trip travel times, from stack base to PTA

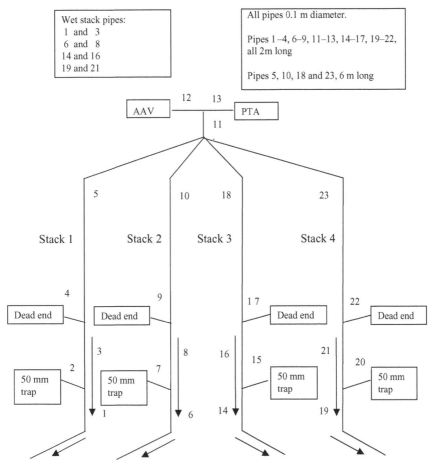

Figure 6.1. Four-stack building drainage and vent system to demonstrate the viability of a sealed building system

of approximately 0.08 seconds and from stack base to stack base of approximately 0.15 seconds.

In order to simplify the output from the simulation, no local trap seal protection is included – for example the traps could be fitted with either, or both, an AAV and PTA as examples of Active Control. Traditional networks would of course include passive venting where separate vent stacks would be provided to atmosphere.

Ideally the four sewer connections shown should be to separate collection drains so that diversity in the sewer network also acts to aid system self-venting. In a complex building this requirement would not be arduous and would in all probability be the norm.

It is stressed that the network illustrated is representative of complex building drainage networks. The AIRNET simulation will allow a range of appliance discharge and sewer imposed transient conditions to be investigated.

The following appliance discharges and imposed sewer transients were considered:

1   w.c. discharges to stacks 1 to 3 over a period 1 to 6 seconds and a separate w.c. discharge to stack 4 between 2 and 7 seconds;
2   a minimum water flow in each stack continues throughout the simulation, set at 0.1 litres/second, to represent trailing water following multiple appliance discharges;
3   a stack base surcharge event is assumed to occur in stack 1 between 2.5 and 3.0 seconds;
4   sequential sewer transients imposed at the base of each stack in turn for 1.5 seconds from 12 to 18 seconds.

The application of the AIRNET simulation will demonstrate the efficacy of both the concept of active surge control and inter-stack venting in enabling the system to be sealed, i.e. to have no high-level roof penetrations and no vent stacks open to atmosphere outside the building envelope.

The imposed water flows within the network are based on 'real' system values, being representative of current w.c. discharge characteristics in terms of peak flow, 2 litres/second, overall volume, 6 litres, and duration, 6 seconds. Figure 6.2 illustrates the w.c. discharges described above. The sewer transients at 30 mm water gauge are representative but not excessive.

It should be noted that heights for the system stacks are measured positive upwards from the stack base in each case. This implies that entrained airflow towards the stack base is negative. Airflow entering the network from any AAVs installed will therefore be indicated as negative. Airflow exiting the network to the sewer connection will be negative. Airflow entering the network from the sewer connection or induced to flow up any stack will be positive. Water downflow in a vertical is however regarded as positive. Observing these conventions will allow the following simulation to be better understood.

The boundary conditions for the system illustrated have been covered in Chapter 3, however the detail of the applicable boundary condition at the base

Applied annular downflow in each of the four system stacks. Note simultaneous flows in pipes 1, 6 and 14 and overlapping discharge, delayed by 1 second, in pipe 19.

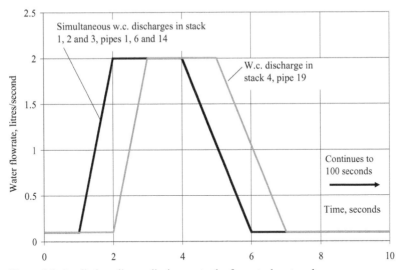

Figure 6.2 Applied appliance discharges to the four-stack network

of each stack requires some further explanation, Figure 6.3. Under normal circumstances the annular downflow from the vertical stack passes through the stack base boundary formed by a water curtain that forms across the entry to the horizontal drain as shown, provided the downflow is sufficient. This generates the expected positive back pressure encountered in the traditional stack pressure regime. The boundary condition to be solved with the appropriate C⁻ characteristic is the pressure loss expression for the water curtain including its water curtain loss coefficient. The water curtain loss coefficient is an experimentally based term built into the simulation.

Under surcharge conditions the water curtain is sufficient to inhibit air movement into the horizontal drain so that the boundary condition becomes zero airflow velocity, generating a Joukowsky air pressure transient.

If the prevailing pressure regime in the network is such that airflow is entrained into a stack from the horizontal drain then any water film resistance will operate to oppose the reverse flow into the stack, resulting in a pressure in the stack lower than that in the horizontal drain. In the current example this may occur if some of the vertical stacks are subject to annular water downflow following appliance discharge while others still only carry a trickle flow or are dry. As shown earlier it is normal to express the pressure loss expressions in terms of (velocity | velocity |) rather than velocity squared to ensure that the loss always opposes the flow and the inclusion of this form of the loss equation in the stack base boundary allows air movement in either direction to be catered for. Depending on the transient response of the system as a whole, a reduced reversed airflow may still occur in a stack carrying its own appliance discharge,

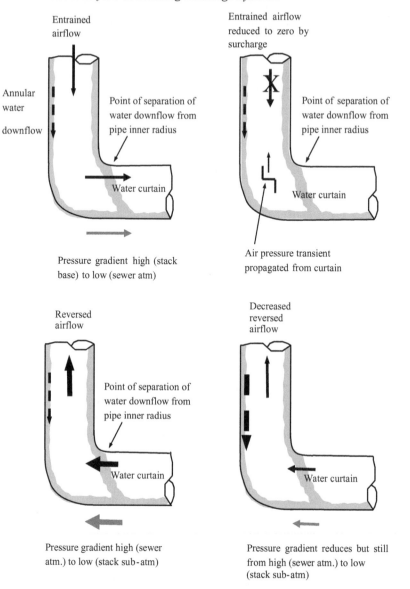

Figure 6.3 Conditions at the base of each stack under normal flow conditions, surcharge and sewer pressurisation

and the characteristics modelling of the annular water flow to entrained air core boundary is sufficiently robust to cope with this eventuality, as will be demonstrated later.

Figure 6.3 illustrates these amendments to the stack base boundary condition. It will be seen that each has a role in the simulation of the overall system response to flow condition changes.

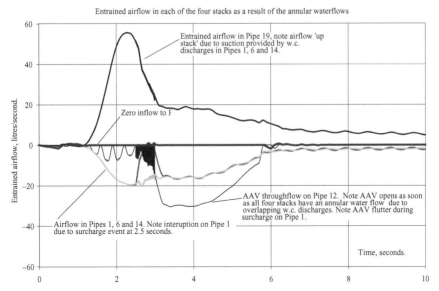

Figure 6.4 Entrained airflows during appliance discharge

As a result of the appliance discharges listed it will be seen from Figure 6.4 that entrained air downflows are established in pipes 1, 6 and 14 as expected. However the entrained airflow in pipe 19 is into the network from the sewer. Initially, as there is only a trickle water flow in pipe 19, the entrained airflow in pipe 19 due to the w.c. discharges already being carried by pipes 1, 6 and 14, is reversed, i.e. up the stack, and contributes to the entrained airflow demand in pipes 1, 6 and 14. The AAV on pipe 12 also contributes but initially this is a small proportion of the required airflow and the AAV flutters in response to local pressure conditions.

Following the w.c. discharge to stack 4 that establishes a water downflow in pipe 19 from 2 seconds onwards, the reversed airflow initially established diminishes due to the traction applied by the falling water film in that pipe. However the suction pressures developed in the other three stacks still result in a continuing but reduced reversed airflow in pipe 19. As the water downflow in pipe 19 reaches its maximum value from 3 seconds onwards, the AAV on pipe 12 opens fully and an increased airflow from this source may be identified. The flutter stage is replaced by a fully open period from 3.5 to 5.5 seconds.

Figure 6.5 illustrates the air pressure profile from the stack base in both stacks 1 and 4 at 2.5 seconds into the simulation. The air pressure in stack 4 demonstrates a pressure gradient compatible with the reversed airflow mentioned above. The air pressure profile in stack 1 is typical for a stack carrying an annular water downflow and demonstrates the establishment of a positive backpressure due to the water curtain at the base of the stack.

Following the completion of w.c. discharges, the airflows will naturally attenuate over a period of time based on the frictional resistance of the network.

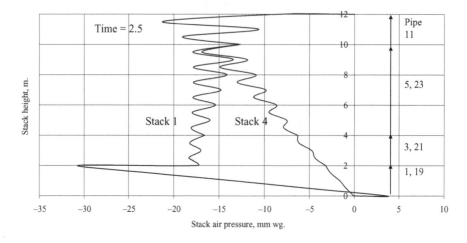

Air pressure profile from stack base to common pipe 11 and AAV/PTA junction in stacks 1 and 4 at 2.5 seconds into the simulation.

Figure 6.5 Air pressure profile in stacks 1 and 4 illustrating the pressure gradient driving the reversed airflow in pipe 19

As a 'trickle' flow is assumed to continue in each stack, the rate of attenuation of the entrained airflows is low.

The initial collapsed volume of the PTA installed on pipe 13 was 0.4 litres, with a fully expanded volume of 40 litres, however due to its small initial volume it may be regarded as collapsed during this phase of the simulation.

Figure 6.6 illustrates the effect of a surcharge at the base of stack 1, pipe 1 at 2.5 to 3.0 seconds. The entrained airflow in pipe 1 reduces to zero at the stack base and a pressure transient is generated within that stack. The impact of this transient will also be seen later in a discussion of the trap seal responses for the network.

It will also be seen, Figure 6.6, that the predicted pressure at the base of pipes 1, 6 and 14 conform to that normally expected, namely a small positive back pressure as the entrained air is forced through the water curtain at the base of the stack and into the sewer. In the case of stack 4, pipe 19, the reversed airflow drawn into the stack demonstrates a pressure drop as it traverses the water curtain present at that stack base.

The AIRNET simulation allows the air pressure profiles up stack 1 to be modelled during and following the surcharge illustrated in Figure 6.6. Figures 6.7 and 6.8 illustrate the air pressure profiles in the stack from 2.5 to 3.0 seconds, the increasing and decreasing phases of the transient propagation being presented sequentially. The traces illustrate the propagation of the positive transient up the stack as well as the pressure oscillations derived from the reflection of the transient at the stack termination at the AAV/PTA junction at the upper end of pipe 11.

Figure 6.9 illustrates the imposition of a series of sequential sewer transients at the base of each stack. Figure 6.10 demonstrates a pattern that is repeated as each stack base is surcharged. It is noteworthy that the simulation accurately indicates that the reversed flow in the affected stack, in this case stack 1, is balanced by the inflow to the PTA and the venting provided by stacks 2–4 leading to diversified sewer connections.

As the positive pressure is imposed at the base of pipe 1 at 12 seconds, airflow is driven up stack 1 towards the PTA connection. However, as the bases of the

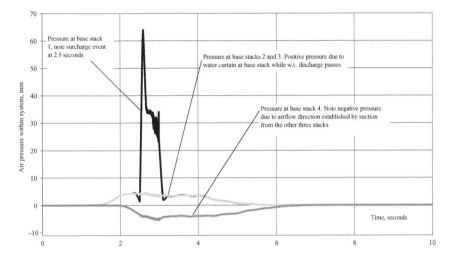

Figure 6.6 Air pressure levels within the network during the w.c. discharge phase of the simulation; note surcharge at base of stack 1, pipe 1 at 2.5 seconds

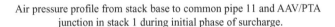

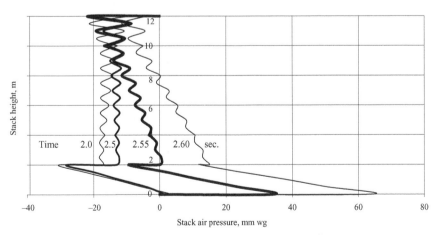

Figure 6.7 Sequential air pressure profiles in stack 1 during initial phase of stack base surcharge

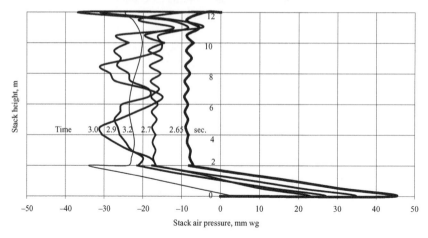

Figure 6.8 Sequential air pressure profiles in stack 1 during final phase of stack base surcharge

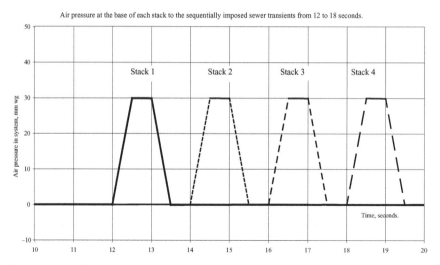

Figure 6.9 Applied sewer pressure transients at each stack base

other stacks have not yet had positive sewer pressure levels imposed, a secondary airflow path is established downwards to the sewer connection in each of stacks 2 to 4, as shown by the negative airflows in Figure 6.10.

As the imposed transient abates, so the reversed flow reduces and the PTA discharges air to the network, again demonstrated by the simulation, Figure 6.10. This pattern repeats as each of the stacks is subjected to a sewer transient. Diversity implies that simultaneous sewer transient imposition would not be a

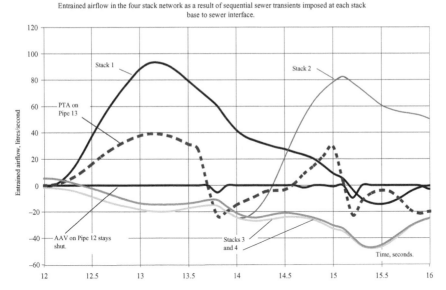

Figure 6.10 Entrained airflows as a result of sewer imposed pressure transients

likely condition and one that would be prudently avoided by ensuring connection to a range of sewer outlets – not an issue for a large complex building.

Figure 6.11 illustrates the repeating sequence of airflow and PTA expansions that follow a series of sewer imposed stack based pressure surges. These figures confirm the balance of flow result discussed for the first stack base imposed pressure, Figure 6.10.

Figure 6.12 illustrates typical air pressure profiles in stacks 1 and 2 during the sewer transient propagation in stack 2 at 15 seconds into the simulation. The pressure gradient in stack 2 confirms the airflow direction up the stack towards the AAV/PTA junction. It will be seen that pressure continues to decrease down stack 1 until it recovers in the lower stack, pipes 1 and 3, due to the effect of the continuing waterflow in those pipes.

The PTA installation reacts to the sewer transients by absorbing airflow. The PTA will expand until the accumulated air inflow reaches its assumed 40 litre volume. At that point the PTA will pressurise and will assist the airflow out of the network via the stacks unaffected by the imposed positive sewer transient. Note that as the sewer transient is applied sequentially from stack 1 to 4 this pattern is repeated. The volume of the high level PTA, together with any others introduced into a more complex network, could be adapted to ensure that no system pressurisation occurred.

Figure 6.13 illustrates the airflow absorbed by the PTA during the sewer transient phase of the simulation. The effect of sequential transients at each of the stacks is identifiable as the PTA volume decreases between transients due to the entrained airflow maintained by the residual water flows in each stack.

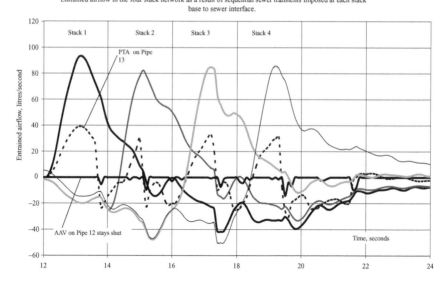

Figure 6.11 Entrained airflows as a result of sewer imposed pressure transients applied to each stack sequentially

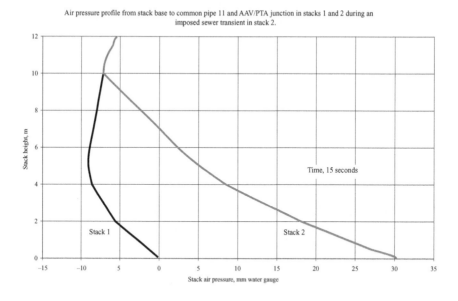

Figure 6.12 Air pressure profiles in stacks 1 and 2 during the sewer imposed transient in stack 2, 15 seconds into the simulation

   It will be seen from Figure 6.13 that the AAV on pipe 12 opens momentarily and this is due to the pressure transient oscillations predicted within the network, as illustrated in Figure 6.14. In practice the inflow to the network from such predicted opening of an AAV is likely to be negligible – indeed in practice it is likely that the AAV would remain shut.

   The appliance traps connected to the network monitor and respond to the local branch air pressures. AIRNET provides a simulation of trap seal deflection, as well as final retention. Figures 6.15 to 6.19 presents trap seal oscillations for one trap on each of the four stacks.

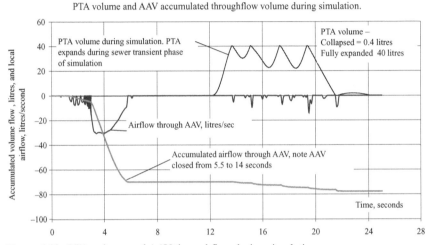

Figure 6.13. PTA volume and AAV throughflow during simulation

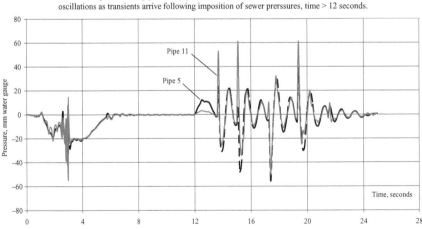

Figure 6.14. Pressure oscillations close to the PTA/AAV junction explaining momentary opening of the AAV illustrated in Figure 6.13

It will be seen from Figure 6.15 that the trap loses seal by induced siphonage during the w.c. discharge as the pressure in the stack falls. The surcharge of the stack base at 2.5 seconds introduces oscillations as the positive pressure transient arrives, causing a reversal of the trap seal water column deflection. Water is forced

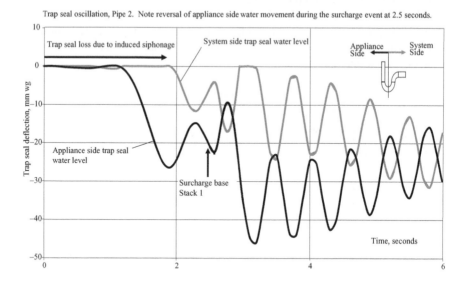

Figure 6.15 Trap seal deflections, trap 2, following initial w.c. discharge and the surcharge of stack 1 at 2.5 seconds

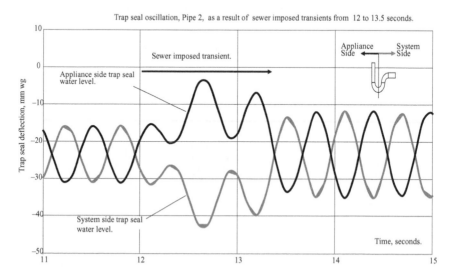

Figure 6.16 Trap seal deflections, trap 2, following the imposition of a sewer transient from 12 to 13.5 seconds

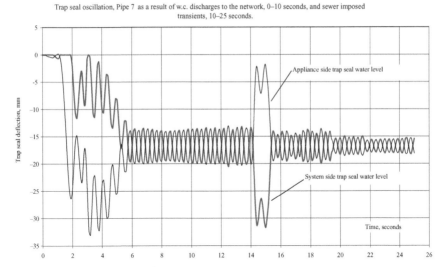

Trap seal oscillation, Pipe 7 as a result of w.c. discharges to the network, 0–10 seconds, and sewer imposed transients, 10–25 seconds.

Figure 6.17 Trap seal oscillation, trap 7

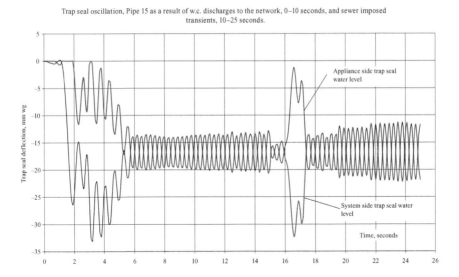

Trap seal oscillation, Pipe 15 as a result of w.c. discharges to the network, 0–10 seconds, and sewer imposed transients, 10–25 seconds.

Figure 6.18 Trap seal oscillation, trap 15

up into the appliance side of the trap. The trap oscillations abate following the cessation of water downflow in stack 1.

Figure 6.16 represents the later imposition of a sewer transient at 12 seconds as the water surface level rises in the appliance side of the trap. A more severe transient could have resulted in 'bubbling through' at this stage if the trap system side water surface level fell below –50 mm.

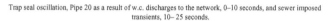

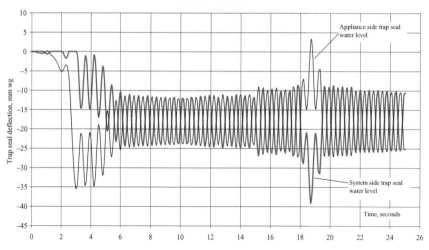

Figure 6.19 Trap seal oscillation, trap 20

The trap seal oscillations for traps on pipes 7 and 15, Figures 6.17 and 6.18, are identical to each other until the sequential imposition of sewer transients at 14 and 16 seconds. Note that the surcharge in pipe 1 does not affect these traps as they are remote from the base of stack 1. The trap on pipe 20, Figure 6.19, displays a later initial reduction in pressure due to the delay in applied water downflow. The sewer transient in pipe 19 is seen as it affects this trap at around 18 seconds.

As a result of the pressure transients arriving at each trap during the simulation there will be a loss of trap seal water. This overall effect results in each trap displaying an individual water seal retention that depends entirely on the usage of the network. Note that the traps on pipes 7 and 15 effectively were exposed to the same levels of transient pressure despite the time difference in arrival of the sewer transients. Further trap seal protection could have been provided by introducing Active Control, in the form of additional AAVs and PTAs adjacent to each trap considered, however this was not demonstrated in order to retain a degree of simplicity in the predicted simulation.

The simulation presented confirms that a sealed building drainage system utilising active transient control would be a viable design option. A sealed building drainage system would offer the following advantages:

- system security would be immeasurably enhanced as all high-level open system terminations would be redundant;
- system complexity would be reduced while system predictability would increase;
- space and material savings within the construction phase of any installation.

These benefits would be realised provided that active transient control and suppression was incorporated into the design in the form of both air admittance valves (AAVs) to suppress negative transients and variable volume containment devices (PTAs) to control positive transients. The diversity inherent in the operation of both building drainage and vent systems and the sewers connected to the building have a role in providing interconnected relief paths as part of the system solution.

The AIRNET simulation has provided output consistent with expectations for the operation of the sealed system studied. The accuracy of the AIRNET simulation in other recent applications, including the accurate corroboration of the SARS spread mechanism within the Amoy Gardens complex in Hong Kong in 2003, provides a confidence level in the results presented.

6.3   Application of the sealed building methodology to real buildings

While the original impetus for the sealed building approach to the application of Active Control to building drainage and vent system design was predicated on the need to cater for security sensitive buildings, the first application of the methodology demonstrated by the simulations presented was to the O_2 'Dome' events venue in East London. The transformation of the O_2 from a 'millennium folly' to a highly successful events venue involved a major refurbishment programme as well as the construction of several 'independent' structures within the original dome. Each of these structures naturally required the provision of utility services, including drainage and vent systems and would normally have required venting to the external environment – in this case the air volume beneath the enclosing dome roof. The local authority ruled that this was unacceptable and hence a solution was proposed to allow the drainage networks required to be designed without external venting of the 'normal' kind.

Figures 6.20 and 6.21 illustrate the form of the overall building. The extensive site is served by several connections to the local area sewer network, Figure 6.22, providing the diversity required to allow consideration of a sealed system, as demonstrated in Figure 6.1.

The provision of multiple sewer connections delivers the diversity criterion for the introduction of a sealed Active Control system and therefore it was possible to proceed to the consideration of the requirements for a successful Active Control solution.

It will be understood from the preceeding simulations that two distinct transient propagation conditions need to be addressed, namely:

1   Negative transient propagation resulting from increases in annular downflow as multiple appliances discharge to the system. It has been shown that these transients may be controllable by introduction of local inwards relief valves that provide the necessary air inflow to the network. The introduction of air admittance valves, both locally to an appliance and at the 'top' of any stack,

Figure 6.20 Aerial view of the O₂ Dome, London (reproduced under Creative Commons License Share Alike 2.5, photo taken by Wikipedia user Debot)

Figure 6.21 O₂ Dome viewed from the Thames (courtesy of A.R. Pingstone, this photo is in the public domain)

has been shown to be a satisfactory way to control and suppress negative transient propagation and hence provide protection for the installed appliance water trap seals without recourse to external venting to atmosphere.

2   Positive transients are generated whenever there is an interruption to the entrained airflow path and can arise from a surcharge either at the base of a vertical stack or at a stack offset. (For this reason good design should wherever possible avoid offsets in the system vertical stacks, although local protection may be applied by introducing an AAV immediately below the offset and a variable volume containment device to provide positive transient attenuation immediately above the offset.) It has been demonstrated in Chapter 5 that a variable volume containment device, or positive air pressure attenuator (PTA), may be used successfully to control and suppress positive transients. Again the introduction of PTA units locally to appliances and at the top of any system vertical stacks will control and suppress transient propagation and provide the required trap seal protection.

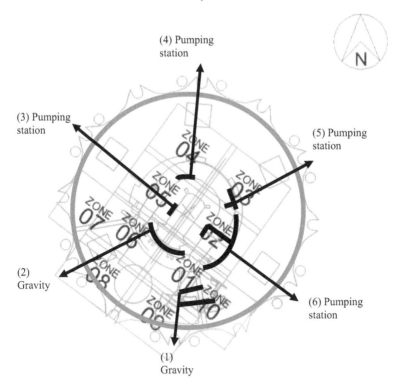

Figure 6.22 Schematic of local area sewer connections to the O_2 Dome

3   In addition to the system generated transients discussed above, the system will also be subject to sewer generated transients, possibly as a result of a remote pump start-up or surcharge of the sewer network. This eventuality is catered for by a combination of the Active Control strategy described, Figure 6.23, and the provision of diversified multiple connections to several local sewer connections that statistically are unlikely to be simultaneously surcharged.

Since opening, some 2.5 million people have used the O_2 complex with no drainage system failures reported to the local authority (Greenwich) due to the choice of an Active Control sealed building system White and Chang (2009).

6.4   AIRNET simulation of the Amoy Gardens SARS virus cross-contamination spread in 2003

Earlier chapters have emphasised the role of the appliance trap seal in preventing cross-contamination of habitable space. The Method of Characteristics low amplitude air pressure transient simulation, AIRNET, has been introduced and shown to successfully model the effects of transient propagation. However, trap seal

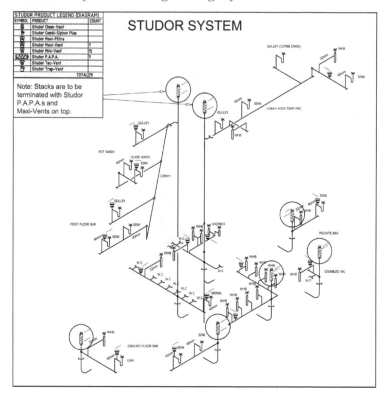

Figure 6.23 Typical detail of the sealed building Active Control solution applied at the O₂ Dome, illustrating the combined use of AAV and the positive transient attenuator P.A.P.A.™ devices to control and suppress transient propagation

loss may be caused by both transient action, exacerbated by any underestimation of the likely system usage, and by poor maintenance or unauthorised modifications to the drainage network, Swaffield and Jack (2004). All of these causes contributed to the involvement of the Amoy Gardens drainage system in the spread of the SARS virus in 2003. This identification of the under-performance of the building drainage and ventilation system as a significant contributor to the transmission of the SARS virus in the Amoy Gardens outbreak, WHO (2003, 2004), not only focused attention upon society's fundamental prerequisite for adequate sanitation, but also initiated a re-examination of the methods adopted to ensure implementation and sustainability of appropriate design methods.

In comparison to many less well-developed communities, Hong Kong has enjoyed the benefits of a relatively sophisticated design approach to building drainage. Recommendations for the safe and effective installation and operation of systems are detailed in the Hong Kong government's Building Regulation document CAP123I (Standards of Sanitary Fitments, Plumbing, Drainage Works and Latrines) and, as with most documents of this kind, the process of review remains ongoing with the aim of ensuring best practice to accommodate changes in building and system design and use. However, the outbreak of SARS in the Amoy Gardens housing complex in March 2003 resulted not only in a re-examination of Regulation CAP123I by Hong Kong government officials and affiliated organisations, but also, in a wider context, an examination of the need for international plumbing guidelines.

The initial WHO investigation of the Amoy Gardens outbreak emphasised the failure of the plumbing system:

> droplets originating from virus-rich excreta ... re-entered into residents' apartments via sewage and drainage systems where there were strong upwards air flows, inadequate 'traps' and non-functional water seals.
>
> Press Release WHO/780, WHO (2003)

The initial studies concluded that the virus was transmitted into habitable space via open floor drain traps. In total there were 321 infected cases in Amoy Gardens, leading to 42 deaths (Hung et al. 2006).

Later studies identified the operation of bathroom fans as a contributory cause that enhanced the cross-contamination via the dry floor trap, Yu et al. (2004), Abraham (2005), Chan et al. (2005). Simulation studies by Swaffield (2005), Jack (2006) and Jack, Cheng and Lu (2006) confirmed the role of both the dry trap and the fan operation in establishing an air exchange between habitable space and the drainage vertical stacks carrying a virus laden entrained airflow.

With similarities to practices widely adopted throughout Europe, the drainage and ventilation pipework design used in Amoy Gardens at the time of the outbreak was based around the use of separate discharge and ventilation stacks, representing an 'abbreviated' form of the traditional 'one-pipe' system. Typically, 'one-pipe' systems use two full-height stacks (one for both 'grey' and 'black' water, and the other for ventilation purposes only), and introduce loop vents between each appliance and the ventilation stack. In the Amoy Gardens case, the 'abbreviated' system adopted loop vents for w.c.s only, Figure 6.24, thus providing an intermediate degree of venting for the appliance type demonstrating the most pronounced variation in discharge profile.

Figures 6.25, 6.26 and 6.27 illustrate the Amoy Gardens building layout. It is stressed that the total population of this housing complex approached 19 000 at the time of the outbreak. An important feature of the design is the services re-entry ducts set into the façade of the building – these are some 7 m in depth and 3 m wide and carry all the drainage services external to the building. Bathroom windows also open onto these deep channels that run the full height of the building and act as a conduit for any buoyancy driven flows emanating from the apartment windows.

As will be seen from the schematic of the installed drainage system, Figure 6.24, the prevention of cross-contamination of the habitable space depends upon the integrity of the trap seals in each apartment. It is well known that the primary purpose of the trap seal is to provide a physical separation between the miasma present within the discharge stack and the habitable space occupied by the building user. Trap seals have been in common use since the 1800s, and thereafter, during the early 1900s, there was recognition of the fact that appliance discharge flows generated air pressure transients that, depending upon their magnitude, had the potential to affect retention levels. This acknowledgement prompted much of the early research in this area, and ultimately resulted in the design evolution that yielded the systems commonly used today, where initial trap seal depth is normally set such that a minimum retention level is sustained following the application of predetermined acceptable pressure excursions from atmosphere (both positive and negative).

Figure 6.28 illustrates the likely transmission route for the SARS virus. The initial infected occupant was identified within the complex. Due to the nature of the infection involving a diarrhoeal phase the w.c. appliance discharge to the system vertical stack would have been a fluid mix of water and infected faecal particles. As the floor drains in bathrooms within the complex were often dry due to internal building temperatures, lack of maintenance and unauthorised modifications, a route existed for the passage of contaminated air from the vertical stack into the

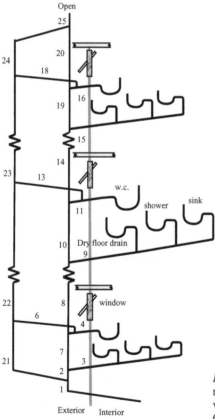

Figure 6.24 Schematic diagram representative of the discharge and ventilation pipework installed in Amoy Gardens, Hong Kong at the time of the SARS outbreak in March 2003

Figure 6.25 Street-level view of the Amoy Gardens complex that housed a population of some 19 000 in 2003 (courtesy of Henry Hung, Ridgid Plumbing Ltd, Hong Kong)

Figure 6.26 Extent of the re-entrant over the full height of the building (courtesy of Henry Hung, Ridgid Plumbing Ltd, Hong Kong)

Figure 6.27 Details of the external services and window openings into the re-entrant on the Amoy Gardens façade. Courtesy of Henry Hung, Ridgid Plumbing Ltd, Hong Kong

habitable space driven by both the transient pressures prevailing in the stack and by the suction provided by any bathroom fan operation, as illustrated.

Figure 6.28 also illustrates the drainage network at Amoy Gardens. In common with other air pressure transient propagation in building drainage networks, as the annular water flows change with time so air pressure transients are propagated throughout the network to communicate the changed system operating conditions. Rapid increases in annular downflow lead to severe negative transient propagation – severe in this application being taken to indicate pressures in the range 50–100 mm of water gauge that would be sufficient to deplete any appliance trap seal. The prevention of trap seal loss is essential to prevent the ingress of sewer gases, contaminated air or odours into habitable space.

If the air path down the duct is obstructed, as may be the case if the water downflow is sufficient to cause surcharging at the base of the stack, then the boundary condition at the base of the stack becomes zero airflow velocity. If this closure of the air path is deemed instantaneous, or less than the period calculated from stack height and air acoustic velocity, then a Joukowsky (1900) style pressure rise is experienced that travels up the stack and interacts with each system boundary. Positive air pressure transients displace the appliance trap seals and may lead to passage of contaminated air through the trap into habitable space.

It is interesting to note that in this application the instantaneous stoppage of an airflow at 1 m/s will generate a pressure transient roughly equal to 40 mm of water gauge, an indication of the relative scale of the transient condition being studied.

This analysis of the transient behaviour of building drainage systems when subject to unsteady annular water downflows and stack surcharge was fundamental to an understanding of the spread of the SARS infection within multi-storey

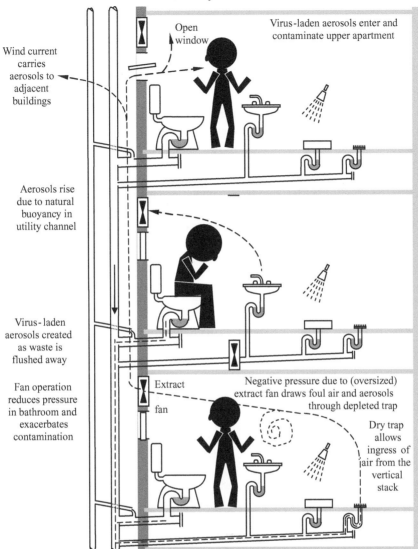

Figure 6.28 Agreed cross-contamination route followed by the SARS virus; the overall spread was recognised as due to a mixture of causes including dry floor traps, bathroom fan operation and natural convective currents within the façade re-entrants

housing in Hong Kong in 2003. Figure 6.28 illustrated the sequence of events that led to contaminated air entering bathroom and shower cubicles within the complex. Due to lack of maintenance the floor drains in several bathrooms were depleted and acted as open terminations for the system, allowing contaminated air to enter the shower cubicle when either the shower room fan was activated or whenever a positive air pressure transient in the stack generated an airflow through the dry trap.

Figure 6.29 illustrates the positioning of the bathroom floor drain. It will be seen that 'topping up' this trap seal would require an inflow of water at floor level – often historically provided by floor mopping, a cleansing process superseded by the use of hygiene wipes if carried out at all.

Unauthorised modifications to the drainage provision within the complex was also identified by the Hong Kong government study and the WHO report; as illustrated in Figure 6.30 the trap provision on the illustrated sink has been removed.

The Method of Characteristics simulation is capable of representing each of these conditions – the fan case being simply modelled by reducing the ambient air pressure above the dry trap. Figure 6.28 illustrated the mechanism for contamination spread that subsequent to passage via the dry trap also included reintroduction through windows further up the building that was subsequently modelled by Computational Fluid Dynamics (CFD) applications dependent upon the airflow around the building complex.

This description of the transmission route is in agreement with the findings of the initial Hong Kong government investigation, as reported by Hung et al. (2006). Figures 6.31 and 6.32 illustrate the floor layout of apartments in the initially affected Block E and the spread of reported infection that generally confirms the route of the virus spread.

Figure 6.29 Floor drain entry, illustrating the practical difficulty in topping up this trap seal (courtesy of Henry Hung, Ridgid Plumbing Ltd, Hong Kong)

Taken together with the schematic Figure 6.24, Figure 6.28 allows an application of AIRNET to be discussed and the main transient mechanisms to be explained in terms of the entrained airflows present in the system.

Initially the system may be assumed to be generally quiescent, with the ventilation fans illustrated operational, Figure 6.24, and a residual 'trickle' water flow of 0.1 litres/second in the vertical stack providing any airflow entrainment or movement. Figure 6.33 illustrates the air movement within the network as a result of fan operation; the net effect of the dry trap is to allow contaminated air from the vertical stack to pass into habitable space. (This phase has been assumed to last 6 seconds within the simulation in order to minimise the duration of the analysis.)

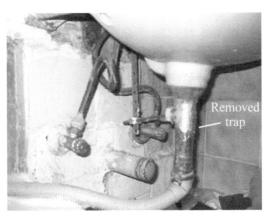

Figure 6.30 Unauthorised modification to a sink where the trap seal protection has just been removed (courtesy of Henry Hung, Ridgid Plumbing Ltd, Hong Kong)

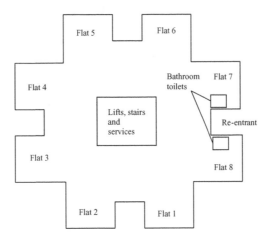

Figure 6.31 Simplified layout of Block E apartments, Amoy Gardens, Hong Kong 2003, (after Hung et al. 2006)

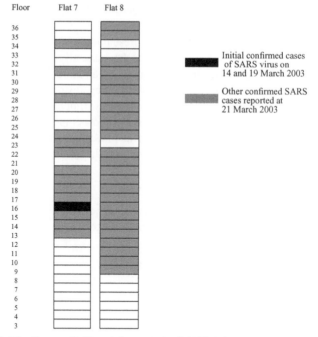

Figure 6.32 Distributon of affected flats at 1 April 2003 (after Hung et al 2006)

Appliance discharge to the stack is illustrated in Figure 6.34, represented by two w.c. discharges between 6–10 seconds and 15–19 seconds. A stack base surcharge is also illustrated in Figure 6.34, between 16.5 and 17.5 seconds.

Figure 6.34 illustrates the effect on entrained airflow as a result of the first w.c. discharge from 6 to 10 seconds. It will be seen that the air entrainment effect of the annular discharge in the vertical stack is sufficient to overcome the fan effect and air is drawn out of the bathroom through the dry trap on pipe 9, Figure 6.24. This increases the airflow in the lower stack section. As the w.c. discharge abates, the fan suction effect is reasserted and the entrained airflow again decreases and air is drawn back into the bathroom. Effectively the combined fan and annular water flow entrainment effect sets up an air exchange between the bathroom and the vertical stack that effectively pumps contaminated air laden with virus into habitable space.

While the fan effect obviously dominates the cross-contamination via the dry trap in the upper floors of the building, a dry trap on a lower floor can be a route for contamination following a stack base surcharge, as shown in Figure 6.35. In this case the reverse airflow generated by the surcharge is sufficient to pump contaminated air into the habitable space through the depleted trap on pipe 3.

The AIRNET simulation provided a tool used to undertake a forensic simulation of the 2003 SARS epidemic cross-contamination in Amoy Gardens, Hong Kong and was used to determine the route by which contaminated air may have entered habitable space in Amoy Gardens. The outcome was supportive of the WHO and local authority investigations of the infection spread.

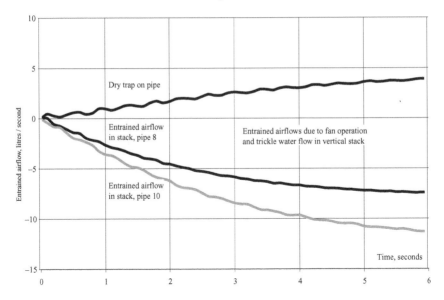

Figure 6.33 Entrained airflow to the bathroom from the drainage stack during a quiescent phase with the fan in operation

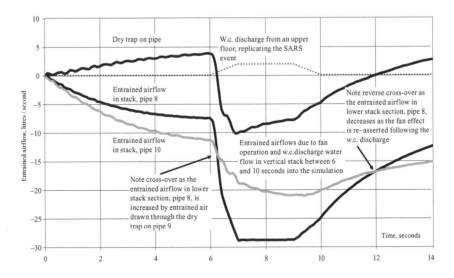

Figure 6.34 Entrained airflow to the bathroom from the drainage stack as a result of w.c. discharge followed by a re-establishment of the fan-induced ingress of contaminated air; this cyclic process establishes an air exchange between the bathroom and the vertical stack

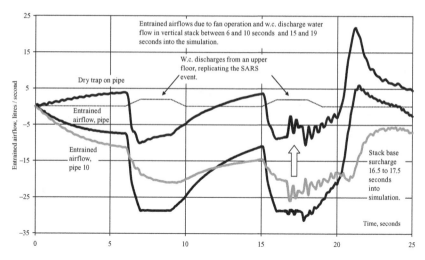

Figure 6.35 Entrained airflow to the bathroom from the drainage stack as a result of w.c. discharge may be strengthened if a stack base surcharge occurs; the resulting positive transient pressure effectively drives contaminated air into the bathrooms at the lower levels of the building without the contribution of fan operation

Thus Figures 6.33 to 6.35 illustrate the Method of Characteristics simulation that predicts contaminated air movement into habitable space through dry traps on each level as a result of cubicle fan operation and following surcharge at the base of the stack. Initial low flow annular downflow generates entrained airflows in the main stack – up to 6 seconds. Fan operation at the pipe 9 level draws an airflow into the cubicle that continues until annular water flow in the stack generates a greater suction and reverses the flow, at 6–10 and 15–19 seconds. Surcharging the base of the stack, at 17 seconds, drives air through pipe 3, connecting the bathroom on that level to the stack, due to the positive surge pressure generated by airflow stoppage. Upper levels may be sufficiently remote from the surcharge event not to be affected; airflow is consistently into the stack through the dry trap on these levels.

The initial response to the dry trap condition was to find a solution that would allow rapid retrofit and avoid the possibility of a recurrence of the trap seal depletion without major alterations to the habitable space. Figure 6.36 illustrates a solution implemented where an additional running trap was added external to the building within the service system carrying re-entrant discussed previously. This trap effectively protects all the appliances, with the exception of the w.c. suite, from cross-contamination from the system's vertical stacks. It will be seen that even if the floor drain trap dries out there is still protection as the running trap will be automatically recharged by any usage of the sink or shower.

Another solution considered at the time, but not implemented, was the replacement of the water trap seals with sheath traps, as discussed in Chapter 5.

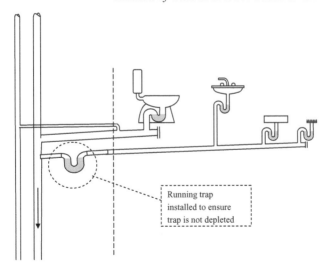

Figure 6.36 Remedial running trap installed in the re-entrant to ensure trap depletion eliminated

These could be used either vertically below any appliance or in a manner analogous to the running trap illustrated.

6.5 Depleted trap seal identification – a response to the SARS outbreak

The loss of a water trap seal is, by definition, a failure in the building drainage system and is typically characterised by the ingress of an unpleasant odour to the habitable space. It was not until the outbreak of the severe acute respiratory syndrome (SARS) virus in 2003 that the potential health risks were first fully realised. Identification of the drainage system as a transmission route of the virus at the high-rise apartment buildings of Amoy Gardens in Hong Kong sent shockwaves around the world. The virus was first introduced to the drainage network by infected residents suffering severe diarrhoea – a symptom of SARS. Contaminated water droplets were then drawn back into the building via dry floor drain traps, aided by the sub-atmospheric pressure generated by the operation of bathroom extract fans which then discharged the contaminated air to atmosphere, further exposing the virus to neighbouring apartments and adjacent buildings. Globally, SARS infected 8089 people and claimed 774 lives. Out of the 321 Amoy Gardens residents infected with the virus, there were 42 fatalities.

The Amoy Gardens SARS outbreak was an important reminder of both the importance of high-quality drainage installation and control of system usage and adaptation. In addition the outbreak highlighted the central importance of the trap seal in preventing cross-infection and placed an emphasis on ensuring that trap seal depletion is eliminated by good design that recognises the role of transient propagation within building drainage and vent systems. However trap seal depletion will occur due to unexpectedly high system usage, maintenance failures or externally imposed pressure transients from the sewer system. Therefore, as a result of the analysis of the route of the SARS infection spread, a research programme was initiated at Heriot Watt University to investigate the practicality of a remote, non-invasive, non-destructive test methodology that would identify the location and frequency of trap seal depletion in complex buildings, Swaffield (2005). The research was a direct result of the SARS investigation and was supported by the UK Engineering and Physical Sciences Research Council as well as an industrial sponsor Studor and supporting organisations, including the World Plumbing Council, IAPMO, SNIPEF, BuroHappold and the Royal Bank of Scotland and therefore represented a concerted effort by the world plumbing community to learn from the SARS outbreak.

The study of low amplitude air pressure transient propagation within building drainage and vent systems, as discussed in this book, has traditionally been concerned with the identification and elimination of pressure transients; however it was clear that the introduction of low amplitude air pressure transients could be utilised to identify the presence of depleted trap seals by utilising the different characteristic reflection coefficients of fully charged and depleted traps discussed in earlier chapters.

The practicalities of current maintenance methods (which rely on cumbersome and time-consuming visual inspections) prevent the regular and routine checking of trap seal status, particularly in large complex buildings. The non-invasive technique described here draws upon a basic understanding of the reflection and transmission of low amplitude air pressure transients in building drainage and vent systems as well as sharing a theoretical underpinning with similar methods proposed to determine leak locations in large-scale water distribution systems, Brunone and Ferrante (2004) and Stephens et al. (2004). The depleted trap identification methodology has a major advantage over these other applications as the location of possible dry traps are known, whereas leaks may occur at any unspecified location in a complex system and therefore it is likely that the drainage application will benefit from this simplification.

The basis of this proposed technique centres on the transient reflection and transmission of pressure waves at network boundaries, either internal or terminal. As an air pressure wave propagates throughout the drainage network it will be reflected and/or transmitted by every system boundary, whether a branch to stack junction, air admittance valve, water trap seal or an open or closed end. The reflection returned from each boundary carries important information regarding its physical characteristics and position within the system. If one of these boundaries is changed then this is shown as a change to the pressure–time history at the time taken for the changed reflection to return to the monitoring location. Therefore, it is possible to assume that the change from a full water trap seal to an empty water trap seal can be detected by the change in transient response at that boundary.

Laboratory tests, as well as four sets of field trials, will be discussed to confirm the efficacy and practicality of this technique. The numerical simulation model,

AIRNET, will be shown to both accurately predict the system response to an applied pressure transient and offer an invaluable pre-test system analysis. Special attention was given to ensuring that the test methodology was non-invasive and did not itself pose a threat to the system integrity.

6.6   Basis for a non-invasive depleted trap seal methodology

There are three main properties of wave propagation in pipes that are intrinsic to the implementation of this transient based technique. These are:

1   the change in reflection coefficient between a full and empty trap seal,;
2   the reflection and transmission of a wave at a junction;
3   the successful determination of the time at which a reflection is returned to source.

The water trap seal is normally located immediately downstream of an appliance and so forms the termination of that branch connection. For a full trap (analogous to a closed end) the termination boundary implies an airflow mean velocity equal to zero and therefore is unable to sustain the fluid flow induced across the approaching pressure wave, positive or negative, and, hence, generates a +1 reflection coefficient, i.e. it brings the flow to rest. For an empty trap (similar to an open end) the arriving wave cannot sustain the incoming pressure transient, again positive or negative, due to the constant ambient pressure at the termination boundary and so a −1 reflection coefficient is generated.

Figure 6.37 demonstrates how the change in termination from a closed to an open end affects the measured transient response for a single pipe system subjected to an applied positive pressure wave.

In both cases the transient propagation monitored is identical up to the arrival of the termination reflection at which point the closed end generates a +1 reflection while the open end returns a −1 reflection. This property facilitates the detection of an empty trap by time analysis of the reflected wave.

When a propagating wave encounters a junction it will be partly reflected back along the pipe from which it came while the remainder will be transmitted to the other pipes in the system. Wave division is dependent upon the system geometry and is defined by the following general expressions which can be used to calculate the reflection and transmission coefficients respectively for a junction of n pipes, see Figure 6.38.

It has been demonstrated in Chapter 2 that the reflection and transmission coefficients expected at any junction of n pipes may be expressed in terms of the incoming transient, here taken as in pipe 1, and the transients propagated in pipes 2 to n and reflected in pipe 1; Equations 6.1 and 6.2 reproduce these earlier relationships:

$$C_R = \frac{A_1 - A_2 - A_3 - ... - A_n}{A_1 + A_2 + A_3 + ... + A_n} \tag{6.1}$$

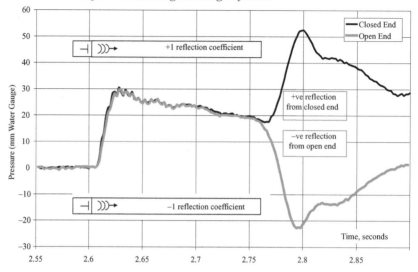

Figure 6.37 Demonstration of the +1/–1 reflection coefficients encountered at a closed and open end termination of a single pipe subject to a positive pressure pulse

$$C_T = \frac{2A_1}{A_1 + A_2 + A_3 + ... + A_n} \qquad (6.2)$$

where A is the cross-sectional area. Note that the above expressions have been simplified as the acoustic wave speed will be sensibly constant for all connecting pipes (temperature dependent between 330 m/s and 340 m/s for the tests reported). It will be shown later that Equations 6.1 and 6.2 are important factors in establishing both the attenuation of a transient wave and the position of the optimum monitoring points within a complex system.

As mentioned previously every boundary that the incident transient wave encounters will generate a reflection that will travel back along the pipe, relaying important system information. The time taken for the wave to travel from the monitoring location to the reflector and be returned to the monitoring station is known as the pipe period, T, and can be defined as:

$$T = \frac{2 \times L}{c} \qquad (6.3)$$

where L is the pipe length from the monitoring location to the boundary and c is the acoustic wave speed. This expression can be used to calculate the location of an empty trap from the arrival time of the returned reflection and allows the pipe periods of all trap seals within the system to be predicted. All important system information will be contained within the maximum pipe period of the system. $(2L_{MAX}/c)$. Hence, for a 40 m high drainage stack the first 0.25 seconds of the pressure–time history will contain all of the characteristic system information.

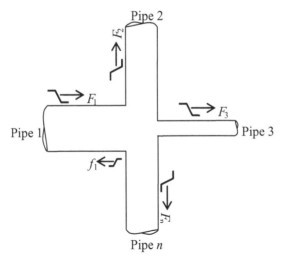

Figure 6.38   General pipe junction with *n* number of pipes

6.7   Laboratory evaluation of the pressure transient identification technique

A laboratory evaluation of the proposed technique was undertaken using a simulated 77 m high vertical drainage stack consisting of 75 mm and 100 mm diameter pipework. Branch connections leading to single appliance trap seals were located at 3.2 m centres representing typical floor levels, Figure 6.39. A pneumatically operated piston, or pressure transient generator (PTG), was located at the simulated stack base which, on activation, would impose a positive transient pulse into the system, the response of which was recorded by a pressure transducer located approximately 1 m from the PTG and connected to a high scan rate data logging system.

Initially all traps were capped off to represent a set of full appliance trap seals – providing the defect-free system for comparison. The system response to the applied single pressure pulse was then monitored for each failure condition by sequentially removing the cap from trap 1 (T1) to trap 14 (T14), thus exposing the trap to atmosphere and simulating a depleted appliance trap seal. Figure 6.40 illustrates the measured system response to the applied pressure transient imposed by the PTG for the defect-free system and selected defective traps at T1, T3 and T12. The pipe periods for each trap location have been calculated using Equation 6.3 and are displayed as vertical lines along the *x*-axis.

The generated transient arrives at the transducer at $t = 1.0$ second. The transient response for the defect-free system shows that the pressure within the pipe continues to rise (to around 100 mm water gauge) during the motion of the PTG until a negative wave is generated between $t = 1.35$ seconds and $t = 1.4$ seconds by

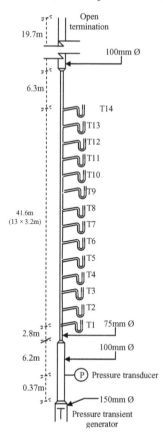

Figure 6.39 Schematic of laboratory test rig

the abrupt cessation of the piston motion. During this period there are no negative reflections returned by any of the traps, thus confirming that all traps are full. The introduction of a dry trap at T1, however, generates a negative reflection with an arrival time of $t = 1.057$ seconds (pipe period of 0.057 seconds) which yields a trap distance of 9 m (consistent with the location of T1, see Figure 6.39). The arrival time of the reflections returned from the depleted traps at T3 and T12 also show good correlation with the predicted pipe periods

6.8 Numerical modelling of the system response

The AIRNET computer simulation, based upon the MoC solution of the St Venant defining equations of unsteady flow, has been used to model the system response of the laboratory system. The action of the PTG at the base of the stack is defined by the boundary equation:

$$u = \frac{\left(x_{piston,t+\Delta t} - x_{piston,t}\right)}{\Delta t} \tag{6.4}$$

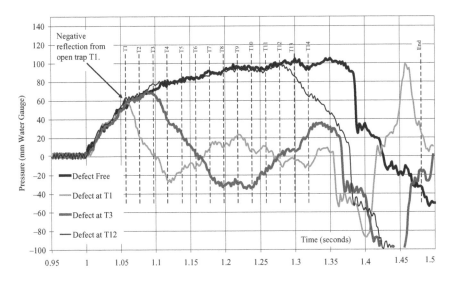

Figure 6.40 Pressure response of the defect-free system compared with those with a depleted trap seal

which may be solved with the available C⁻ characteristic to yield *c* and hence *p* at the piston face. Figure 6.41 shows the simulated transient response for the defect-free system and defective traps at T1 and T12. As was shown with the measured data, the return times of the negative reflections generated by the open traps correspond with those predicted theoretically. This high degree of agreement between the measured and simulated results provides confidence in the Method of Characteristics simulation as a tool to predict system response and as an appraisal of this monitoring technique.

6.9   Selection of an appropriate incident transient wave

In order to measure the transient response it is first necessary to introduce an appropriate transient wave into the system. Initial analysis explored the use of a single positive pressure pulse generated by a pneumatic piston. These preliminary tests were first used to confirm the applicability of the transient technique. Following successful laboratory experiments the technique was then validated with a series of field trials and these are discussed in the next section. Concerns were raised, however, as to the effect of the pulse on the integrity of the system. Applying a pressure pulse into the drainage system could deplete the very trap seals which these tests aim to protect. Further laboratory tests demonstrated that a sinusoidal excitation would offer a 'safe' alternative to the pressure pulse as trap displacement is significantly reduced at frequencies around 10 Hz. The effects of sinusoidal excitation on the displacement of the water trap seal were investigated using the laboratory test rig shown in Figure 6.42.

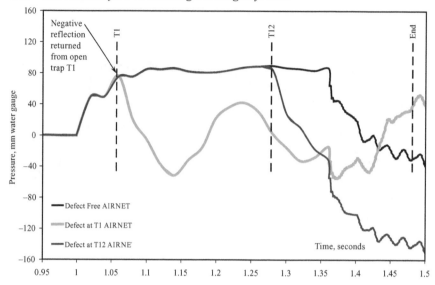

Figure 6.41 AIRNET simulated pressure response of the defect-free system compared with those with a depleted trap seal

Observations confirmed that for an excitation of 1 Hz, the water seal depth oscillated by up to ±6 mm. However, this reduced to only ±1 mm for an excitation of 10 Hz. The AIRNET model, updated to include an unsteady frictional representation using both the Carstens and Roller (1959) and Zilke (1968) analysis for the water column friction in the trap and the airflow within the network, Swaffield (2007), was used to simulate the test conditions. The applied sinusoidal excitation boundary condition is provided by a boundary equation representing the piston velocity at each time step, derived from piston displacement data. The predicted trap seal response to the incoming excitation is shown in Figure 6.43.

It is important to note that the AIRNET prediction of trap seal displacement accurately matches the laboratory observations, corroborated by Beattie (2007). The simulation also confirms that for an excitation of 10 Hz, there is little or no oscillation of the trap seal due to the fluid inertia of the water column. The ±1 mm displacement is observed only as a vibration on the surface of the water seal, confirming that the proposed methodology is truly non-invasive. This study was conducted to verify the laboratory results. Subsequent simulations use closed and open ends in place of the appliance trap seal to reduce simulation run-time.

6.10 Field trial data – the single positive pressure pulse (Dundee)

Initial validation of the proposed test technique using a single positive pressure pulse was provided through controlled field trials on a standard single stack drainage system in an unoccupied 17-storey residential building in Dundee, Scotland, Figure 6.44. The system is shown in Figure 6.45 and consists of a

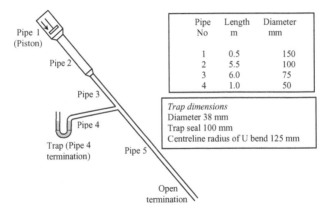

Figure 6.42 Laboratory test rig to investigate the effect of sinusoidal excitation on the deflection of the water trap seal

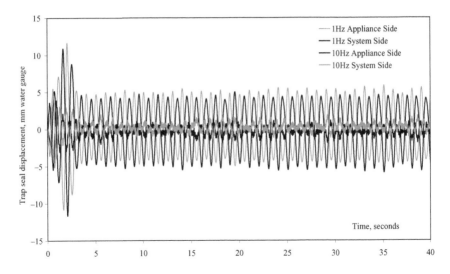

Figure 6.43 Predicted trap seal displacement in response to a 9 mm piston stroke operating at a continuous frequency of 1 Hz and 10 Hz

150 mm diameter cast iron vertical stack with connecting branches serving each appliance on every floor, Figure 6.46.

The pressure transient generator was connected onto the stack via an existing access panel to which a 150 mm inlet branch was connected. The transient response was measured using a single pressure transducer located on the transient inlet branch with a sampling rate of 500 Hz. The defect-free baseline trace was recorded followed by each failure condition by removing the water seal from the w.c. trap on each floor in turn.

Figure 6.44   17-storey residential building in Dundee used for initial field trials

Initial results, Figure 6.47, revealed that the transient response was dominated by a large negative reflection returned at $t = 1.016$ seconds – earlier than any of the predicted trap pipe periods displayed as dashed lines along the x-axis. Depleted traps 2, 6 and 11 were however accurately identified. These results suggested that the large negative reflection was generated from a boundary 2.6 m from the monitoring location. As the incident transient arrived at the branch to stack junction, 2/3 of the wave was transmitted to both the upper and lower sections while 1/3 was reflected back to the source, see Equations 6.1 and 6.2 for a three-pipe junction of equal diameter. As there was nothing in the upper stack section to generate this reflection, the source was identified as the lower stack section connection to the sewer. The large negative reflection from the sewer connection was removed by blocking the lower connection as shown in Figure 6.48. Introducing the closure of one section of the stack also excludes the possibility of 'mirror' reflections from other depleted traps which could confuse the monitored system response if the PTG was connected into the network at an intermediate floor – say floor 9 in Figure 6.45 – where it would be possible to confuse returns from floors 6 and floor 11, say. Initially the sealing off of a stack section was accomplished by the manual insertion of an inflated blocking bladder; however, as discussed later, this became a part of the proposed transient generator equipment by developing a suitable three-port valve. Normal testing would treat each upper/lower stack section sequentially.

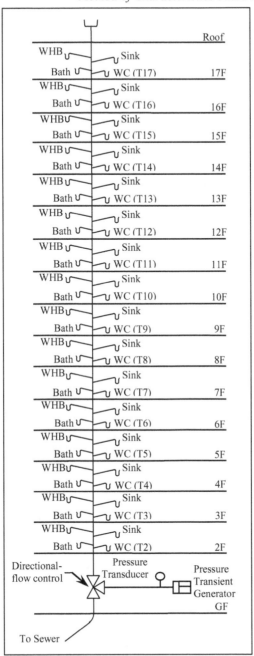

Figure 6.45 Schematic showing the standard single stack drainage system of the 17-storey residential building in Dundee used for initial field trials

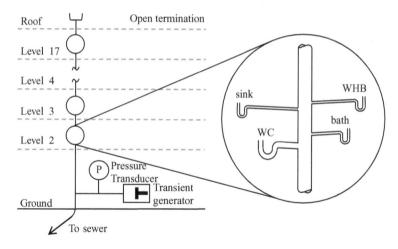

Figure 6.46 Detail of the individual floor drainage network used in the initial Dundee field trials

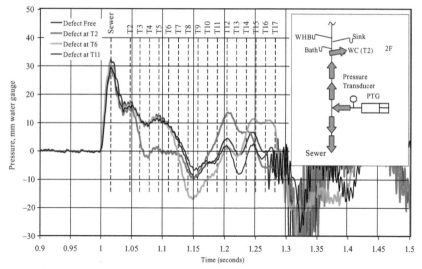

Figure 6.47 System transient response to a single positive pressure pulse with junction open

Preventing transient propagation along both stack sections yields a marked improvement in the magnitude of the propagating wave, Figure 6.49. It can now be seen that by comparing the test trace with the defect-free baseline trace, the reflection time of an empty trap can be clearly identified and accurately corresponds with the predicted trap pipe periods calculated using Equation 6.3. These field trials proved that the detection of empty trap seals through transient analysis was both practical and achievable. The following section examines the non-invasive sinusoidal excitation method.

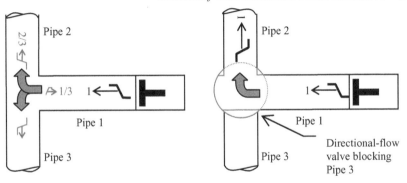

Figure 6.48 Transmission and reflection coefficients for (a) a three-pipe junction, (b) a two-pipe junction created by incorporation of a three-port directional-flow valve

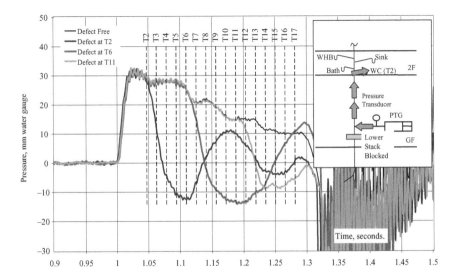

Figure 6.49 System transient response to a single positive pressure pulse incorporating directional-flow valve

6.11 Field trial data – application of a sinusoidal excitation transient at the Arrol Building, Heriot Watt University

Following the successful laboratory and simulation results, the sinusoidal excitation method was evaluated using the drainage system within a department building at Heriot Watt University, Figure 6.50. A schematic of the drainage system is shown in Figure 6.51. Although ranging over only five floors, this drainage system was generally of greater complexity than the 17-storey system studied in Dundee due to the longer branch runs and higher number of connected appliances on each floor.

The 100 mm diameter uPVC vertical stack collects waste from the male toilets and shower room on Levels 3 and 4 as well as the boiler room floor gully on Level 5.

Initial pre-test analysis was carried out using AIRNET to assess the likely system response. Samples of these simulated results are shown in Figure 6.52. A defective boiler room floor gully (T21) can be easily identified by the large negative reflection returned to a pressure transducer on Level 5 at the correct trap pipe period. However, it was discovered that the majority of the defective traps on Levels 3 and 4 did not show the characteristic deviation from the defect-free baseline when monitored on Level 5. Instead, it was difficult to differentiate the defective test trace from that of the defect-free baseline, see the trace for T5 in Figure 6.52, with the Level 5 pressure transducer.

The AIRNET simulation model allows the path of the transient to be traced at each node as it propagates throughout the system. Analysis of the transient at each node showed that it was being largely attenuated by passage through each junction within the system. Taking trap T5 as an example, the transient must first pass nine three-pipe junctions before it arrives at trap T5. As all junctions are, in this case, constructed from equal-diameter pipes (100 mm) the reflection and transmission coefficients are again 1/3 and 2/3 of the incident wave respectively, see Equations 6.1 and 6.2.

The transient arriving at T5 (and, therefore, the negative reflection generated at the open end) is less than 3 per cent of the applied incident wave as the culminative transmission after n junctions is $(C_T)^n$ where C_T is the individual junction transmission coefficient and n is the number of identical junctions passed. The negative reflection is then further attenuated as it travels back along the pipe to the

Plant room on 5th floor utilised to mount transient propagation and monitoring equipment

Figure 6.50 The Heriot Watt University, School of the Built Environment, Arrol Building used for practical evaluation of the technique (courtesy of Heriot Watt University)

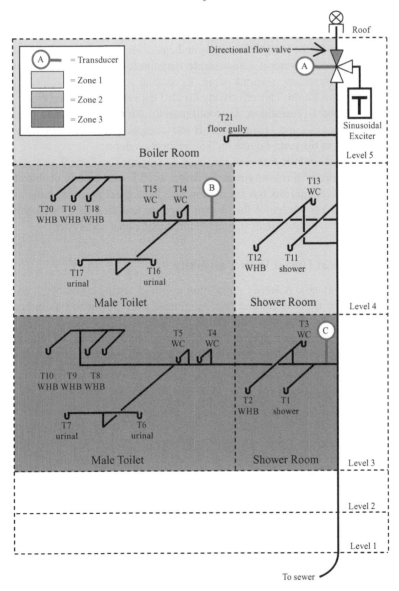

Figure 6.51 Schematic of the drainage system in the Arrol Building, Heriot Watt University, used for practical evaluation of the technique

measurement station where it arrives with magnitude less than 0.1 per cent of the original incident wave, making its detection difficult. It should be noted that for building drainage systems, the main branch connections are generally of a smaller diameter than the main stack and the appliance branch connections are generally of a smaller diameter than the main branch connections. This facilitates a larger wave transmission, as was the case for the system tested in Dundee, where all 17 floors could be adequately monitored from one measurement station.

Interrogation of the AIRNET results indicated locations along the system where the reflections were of a reasonable magnitude to allow their detection. Additional measurement stations were located at each of these positions, improving the resolution. This effectively divided the system into three zones, see Figure 6.51. Zone 1 (Transducer A) would monitor the traps connecting directly into the stack. Zones 2 and 3 (Transducer B and C respectively) would monitor the traps connected to the main branch on their respective floors.

Figure 6.53 shows the transient response as sinulated at Zone 3 and clearly illustrates the improved resolution as the defect at T5 is now clearly visible. This pre-test simulation analysis has been shown to have great practical benefits as it provides prior knowledge of the system response and facilitates the optimum positioning of the measurement stations before testing on a real system.

6.12 Site tests at Heriot Watt University

Following the valuable AIRNET simulation exercise, testing could begin on the real system. The transducers and test equipment were positioned as indicated in Figure 6.51. As predicted, the transient response at Transducer A, Figure 6.54, clearly identifies an empty trap at T21, however, a defect at T5 returns no clear signal. Figure 6.55, however, shows the transient response as measured at Transducer C in Zone 3 where T5 is clearly shown to be empty.

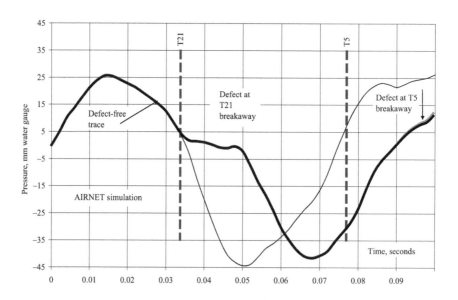

Figure 6.52 Simulated transient response of the system shown in Figure 6.51 to sinusoidal excitation as measured at Transducer A, Zone 1

Importantly, during these field trials trap T21 (the floor gully in the boiler room) dried out daily due to the high temperatures within the boiler room, thus providing a potential route for cross-contamination. Remedial work would be required to remove this hazard by, for example, replacing the gully with a waterless trap. This highlighted yet another practical benefit of this maintenance technique whereby persistent system failures can be identified and 'designed out'.

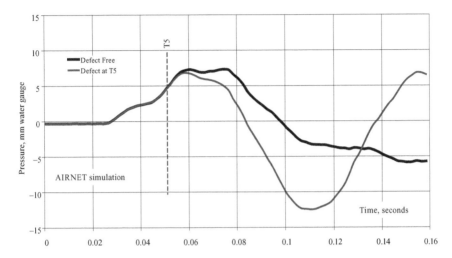

Figure 6.53 Simulated transient response of the system in Figure 6.51 to sinusoidal excitation as measured at Transducer C, Zone 3

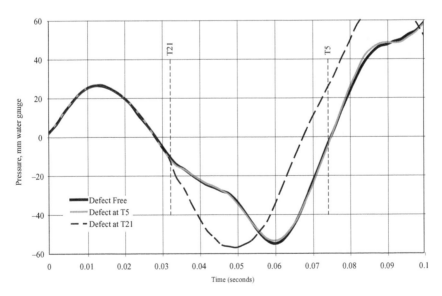

Figure 6.54 Real system transient response of the system in Figure 6.51 to sinusoidal excitation as measured at Transducer A, Zone 1

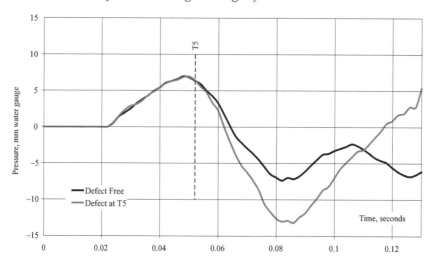

Figure 6.55 Real system transient response of the system in Figure 6.51 to sinusoidal excitation as measured at Transducer C, Zone 3

6.13 Applying the compliance factor to determine reflection arrival time

To determine the location of an open trap seal, an accurate evaluation of the time at which the negative reflection is returned to the measurement station must first be carried out. Once this time is known, the distance to the open trap can be calculated using Equation 6.3. To facilitate automatic recognition of an open trap seal, the absolute compliance factor (ACF) can be applied to calculate the difference between the mean defect-free baseline, mSp_{DF}, and any subsequent test trace, mSp_{TEST}.

$$ACF = \left| mSp_{DF} - mSp_{TEST} \right| \tag{6.5}$$

where m is the measurement station and mSp_{DF}, calculated for N defect-free test traces, is defined by

$$mSp_{DF} = \frac{mSp_1 + mSp_2 + ... + mSp_N}{N} \tag{6.6}$$

The absolute compliance factor can then be compared with a predetermined system parameter to identify the returned reflection time. This system parameter, or maximum deviation (MD), is calculated to find the threshold to which each test trace will be compared by calculating the absolute difference between the mean defect-free pressure trace and each of a sample of N defect-free test traces:

$$MD = \left| mSp_i - mSp_{DF} \right|_{i=1}^{i=N} \tag{6.7}$$

The maximum value of MD becomes the acceptable tolerance for the defect-free system ($mSp_{DF} = mSp_{TEST}$) and is a measure of the spread between the test traces gathered for the identical defect-free system. Any deviation above this tolerance will highlight a failure condition by indicating the presence and location of a new reflection. An example is shown in Figure 6.56. Firstly, the mean defect-free pressure trace, mSp_{DF}, was calculated from 20 defect-free test traces from the system shown in Figure 6.51. The deviation was then calculated for each trace and the maximum deviation established as the acceptable tolerance (in this case 3.96 mm water gauge).

As the text has introduced the concept of Active Control via combinations of AAVs and PTAs it was necessary to consider whether such devices would in any way invalidate the wave reflection approach to depleted trap seal identification. It was found that any reflections from the AAV or PTA boundary conditions would be included automatically in the defect-free trace that is used to derive the values of MD and ACF. Any subsequent trap seal depletion would therefore still be obvious as a deviation from this baseline set of pressure traces. Therefore it is considered that the technique described is applicable to any system where it is possible to evaluate a defect-free baseline.

The compliance factor of the system response with an empty trap (trap T21) can be calculated and compared with the maximum deviation. From Figure 6.56, the point at which the compliance factor exceeds the maximum deviation occurs at 0.034 seconds (the pipe period of trap T21). The value of the maximum deviation is a function of the repeatability of the incident wave. It is envisaged that the maximum deviation will be determined automatically during the test commissioning process.

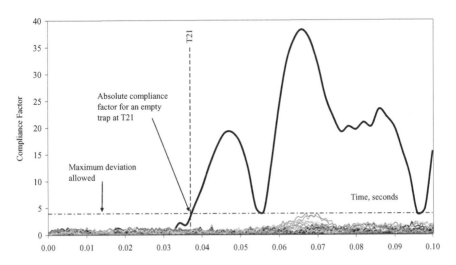

Figure 6.56 Compliance factor identification of the new reflection returned from trap T21 in the system shown in Figure 6.51

6.14 Development of test equipment suitable for use in a building outside the control of the research group

The findings of the field trials facilitated the development of the test technique to include appropriate equipment and computer software in order to perform, record and analyse the system response to an applied transient. The two main requirements are that the route of the traversing transient must be controllable (provided by the directional-flow valve) and that the test must be completely non-invasive (achieved through application of a sinusoidal excitation of around 10 Hz). The required test equipment consists of that shown in Figure 6.57.

The site tests described above had in common a degree of user control as the Dundee building was empty and the Heriot Watt building was under the control of the research team. It was necessary therefore to evaluate the methodology in an operational building not directly controlled by the research team and one where normal commercial activity continued despite the presence of the evaluation team. The Facilities Managers at the Royal Bank of Scotland had been represented on the research group steering panel throughout the development of the methodology described so it was natural for the building chosen for a definitive site evaluation under operational conditions to be an RBS establishment. An office complex in Glasgow was chosen, Figure 6.58, and the equipment was installed in an upper floor plant room. Note that the drainage stack terminates in an air admittance valve rather than a roof penetration.

The building drainage and vent system installed in the RBS building is illustrated in Figure 6.59 and featured six individual stacks serving some 120 appliances.

This multi-stack system consisted of six 100 mm diameter cast iron single stacks, each connecting to a common horizontal drain pipe at high level within the basement, and each terminated with an air admittance valve. Stack 2 and stack 5 served w.c.s only, whilst the other stacks served wash-hand basins only.

One stack was selected to mount the test equipment combination, from where the whole system would be tested to examine the working range of the reflected wave technique. Stack 3 was selected and the equipment was installed within the upper stack section on floor 7. Pre-test analysis was carried out using the Method of Characteristics based computer model, AIRNET, capable of simulating the technique in any building drainage system to allow optimum pressure measurement points to be identified within the system, as previously demonstrated in the Heriot Watt building trials. In addition to the pressure transducer included in the test equipment combination, PT 1, a further transducer was required at the base of each of the remaining stacks, however due to issues of accessibility these transducers were located at floor 2. Each transducer was designated by its *stack number* and *floor number*, i.e. PT 1/2 for a transducer on stack 1 and floor 2, and each trap was designated by its *stack number* and *trap number*, i.e. T 2/12 for a trap on stack 2 and number 12.

To begin the test process the system characteristics in terms of configuration and dimensions were collected, allowing X_D^{true} to be input to the trap recognition

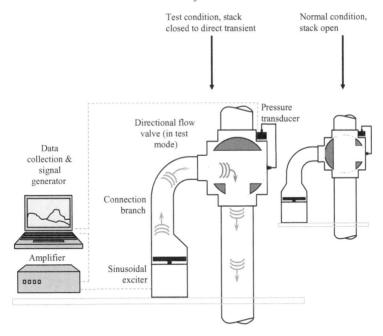

Figure 6.57 Test equipment used to provide a remote and non-invasive transient based technique to identify empty water trap seals

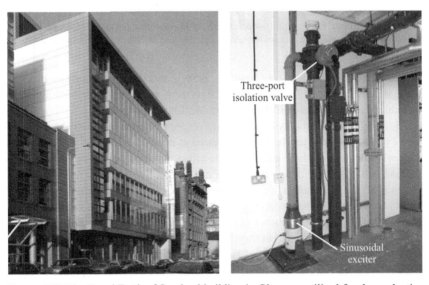

Figure 6.58 The Royal Bank of Scotland building in Glasgow, utilised for the evaluation of the depleted trap identification system in an operational environment; and the detail of the equipment installation, Figure 6.57, in an upper floor plant room (courtesy of RBS)

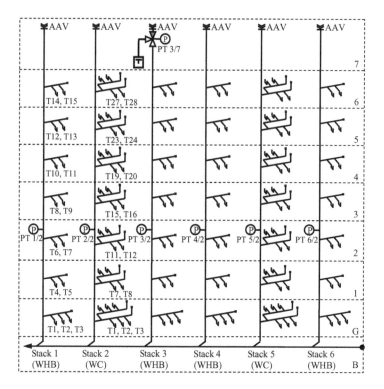

Figure 6.59 Schematic of multi-stack RBS building drainage system used for field investigations showing location of the test equipment combination and pressure transducers

program. Next, the defect-free baseline, mSp_{DF}, and the system threshold, h_{MAX}, were determined from a sample of 20 system pressure responses, again as introduced in the discussion of the Heriot Watt building trials.

The magnitude of the system pressure response measured at PT 1 on stack 3 was 7 to 10 times greater than that measured at floor 2 on any of the other stacks. This was due, as already discussed, to the effect that the system junctions have on the transmitted transient, as defined by Equations 6.1 and 6.2.

Once all the parameters were input to the trap recognition program, depleted traps were created by removing the water seal from a selection of traps on each of the six stacks. Figure 6.60 shows an example of the test system response, obtained while a depleted trap existed at trap T4.12. The test system response is compared with the defect-free baseline and the time at which the two traces diverge to indicate the return time of the reflected wave from the depleted trap. It can be seen that there is a small discrepancy between the perceived return time, t^* (identified by the trap recognition program), and the predicted return time, t (calculated using Equation 6.3 and based on the true trap location, X_D^{true}).

The RBS building represented a complex stack structure with many more junctions than encountered in the previous trial buildings. The effect of a series

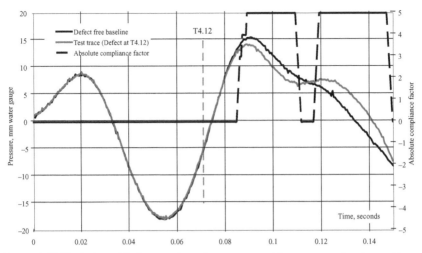

Figure 6.60 Typical graphical output from the trap recognition program showing test trace compared with defect-free baseline and resultant absolute compliance factor, in this case for a defect at trap T4.12

of junctions on the transmitted transient is illustrated clearly in the AIRNET simulation presented in Figure 6.61. While the final transient value is unchanged, multiple junctions introduce multiple bifurcations of the transient followed by multiple re-combinations as a result of branch terminal reflections. This has the effect of delaying the arrival of the transient.

From the RBS building trials, in every test t^* was found to be greater than t, resulting in an *overestimate* of the predicted trap location, X_D^*. Figure 6.62 compares X_D^* with X_D^{true} for every depleted trap tested in the site tests in Dundee, Heriot Watt and Glasgow. It can be seen that the difference increases with trap distance. The magnitude of difference varied between the w.c. stacks and the wash-hand basin stacks, effectively by branch to stack area ratio as the w.c. connections were all 100 mm diameter as were the stacks. This phenomenon confirms the wave delay caused by the pipe junctions illustrated in Figure 6.61. A local junction coefficient based on the transmission and reflection coefficient expressions, Equations 6.1 and 6.2, would explain the delay in the measured return time of the reflected wave and would also explain the variation between the w.c. stack and the wash-hand basin stack in the Glasgow RBS building, owing to the different branch to stack area ratios that are specific to each. The w.c. stack has a 100 mm diameter stack with 100 mm diameter branches, therefore generating reflection and transmission coefficients of 1/3 and 2/3 respectively at each junction, whereas the wash-hand basin stack has a 100 mm diameter stack with 50 mm diameter branches, generating reflection and transmission coefficients of 1/9 and 8/9 respectively. The larger reflection, and hence reduced transmission, generated at each junction in the w.c. stack is responsible for the larger variation between X_D^* and X_D^{true}, in the case of the w.c. stack.

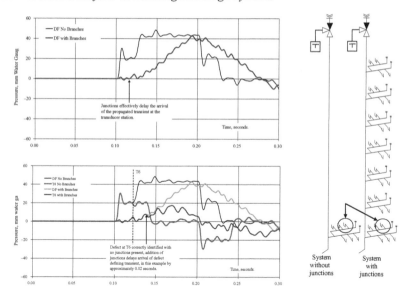

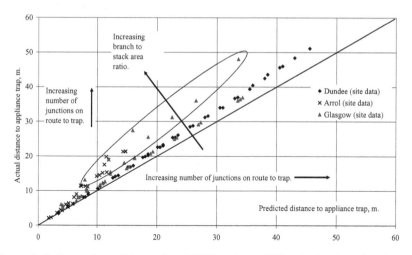

Figure 6.61 Effect of junctions on the AIRNET predictions of transient arrival time and the identification of a depleted trap seal

Figure 6.62 Comparison of the predicted, XD*, and true, XD^{true}, depleted trap locations

Kelly (2009a and b) considered these results and postulated a junction delay factor. To identify the effect of the junctions, C_j would first be quantified as

$$C_j = \frac{X_D * - X_D^{true}}{2n} \tag{6.8}$$

where n is the total number of junctions that the transient wave encounters as it travels from the measurement point to the depleted trap. From the collected data, average values of C_j were calculated for both the w.c. and wash-hand basin stacks and were found to be 0.97 m and 0.21 m respectively. The equivalent pipe length, J_e, attributed to the presence of these junctions in the pipeline can then be determined:

$$J_\varepsilon = 2n \times C_j \qquad (6.9)$$

which can be used to calculate an adjusted predicted trap location, X_D^{adj}, taking account of the local delay coefficient at each junction within the system:

$$X_D^{adj} = X_D * - J_\varepsilon \qquad (6.10)$$

Accuracy can be increased by simply using the reflected wave technique itself to map the trap locations during system calibration to directly determine $t*$ for each possible depleted trap – the PROBE technique. This is made possible as each trap has a fixed location within the system, thus returning a value of $t*$ which is specific to that location. This method removes the need to measure the distance to every trap to determine X_D^{true} and has been found to be highly accurate and repeatable in identifying the correct depleted trap location. Figure 6.63 illustrates this technique for all the site tests undertaken, Kelly (2009a), and compares the PROBE results with the AIRNET simulation that accurately includes the junction effect.

The genesis of the wave reflection technique to identify depleted trap seals was as a response to the Amoy Gardens SARS infection route as identified by the WHO, namely the failure to maintain dry traps. As the development of the methodology progressed it was clear that one major application would be to health care buildings and hospitals in particular. Contacts with the Royal Infirmary

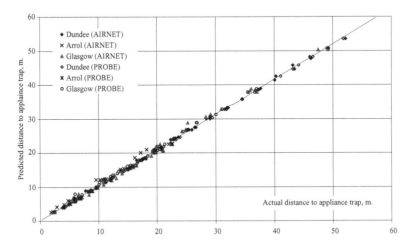

Figure 6.63 Comparison of the true and predicted depleted trap locations recorded during the Dundee, Heriot Watt University Arrol building and Glasgow field trials when using AIRNET and the PROBE method to predict trap locations; these results confirm the ability of AIRNET to model the junction effect time delay

Edinburgh, facilitated through Consort Healthcare and Balfour Beatty Workplace, led to the final site test for the developed equipment. Figure 6.64 illustrates the hospital campus on the south side of Edinburgh. The test equipment, identical to the RBS installation already described, was installed in an upper level plant room above the ward complex.

Figure 6.65 illustrates the vertical stack layout typical for the servicing of the extensive ward areas. Due to the high number of vertical stacks encountered in this site application it was necessary to reconsider the methodology appropriate to deliver the sinusoidal test transient to the stack. It was decided that a single oscillator could be used to serve a multiplicity of vertical stacks via a high level small diameter manifold, each stack being equipped with the motorised three-port valve already discussed, Figure 6.57. This arrangement represents a major improvement in the flexibility of application of the test equipment, Gormley and Hartley (2009).

Figure 6.66 illustrates the familiar divergence between the defect-free system response and the response with a depleted trap on stack 2, Figure 6.65. The test series represented by this figure confirmed the positive outcome of the previous site tests, laboratory investigations and AIRNET simulations. The technique is therefore seen as a valuable contribution to offsetting real concerns as to cross-contamination due to depleted trap seals within the healthcare building environment.

The need to provide a comprehensive monitoring and maintenance regime for the building drainage system became apparent after a defective system was found to have facilitated the spread of the SARS virus in 2003 via depleted trap seals. This chapter has demonstrated that a remote and non-invasive test methodology

Figure 6.64 Aerial view of the Royal Infirmary Edinburgh; the ward accommodation and plant rooms are housed in the right facing arc

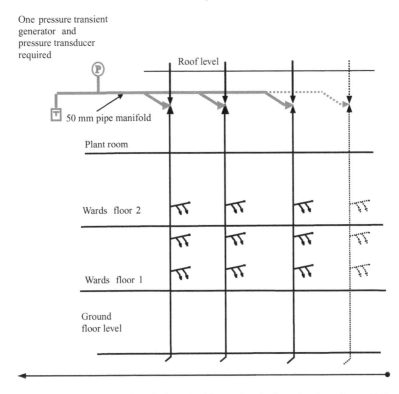

Figure 6.65 Arrangement of vertical stacks, illustrating the introduction of a small diameter manifold to transmit the test signal to each in turn

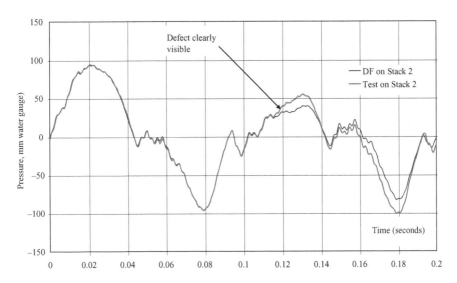

Figure 6.66 Example trace illustrating the response divergence as a result of a dry trap being identified in stack 2, Figure 6.65

is achievable by application of a transient based technique using simple wave propagation theory. Field trials in a 17-storey building in Dundee using a single pressure pulse evaluated the technique and proved that empty traps, returning a negative reflection, could be identified by comparing the test trace with a predetermined defect-free baseline trace. Further trials in a department building at Heriot Watt University provided validation that sinusoidal excitations offer a completely non-invasive test methodology by causing no problematic deflections of the water seal.

A major benefit of this technique is that it only requires prior knowledge of the mean defect-free pressure trace (with which all test traces can then be compared) together with the predicted pipe period of each trap. If there are uncertainties concerning pipe lengths due to lack of system drawings or the inability to manually measure the system, the test can be applied as a 'probe' to determine the reflection times of each trap during the commissioning period. Additional measurement stations may be required to enhance the resolution of the test, particularly in large complex buildings where a higher degree of wave attenuation will occur. Pressure transient simulation of the network has been shown to provide accurate pre-test information of optimal monitoring locations.

The technique provides considerable practical and operational benefits due to the relatively minimal level of equipment required, allowing application in both new and existing systems. There is potential for the technique to be incorporated into a building management system to raise an alarm when a failure condition is identified. This would vastly reduce the maintenance burden currently endured by facility managers and building operators. System designers are also aided, as persistent system failures would be identified allowing remedial work to be undertaken more rapidly and effectively.

In addition to the ability to identify depleted appliance water seal traps it is also clear that the methodology may also be used to identify any unathorised modifications to the network, whether the addition or removal of appliances that require branch connections to the network, as such changes will be readily identifiable as deviations from the stored 'defect-free' or initial status data. It is also clear that the methodology is applicable if the base system includes Active Control elements as the reflections representing these boundary conditions would again be present within the defect-free data.

6.15 Concluding remarks

This chapter has concentrated on the application of low amplitude air pressure transient simulation to a range of 'real' system problems and has demonstrated the role of this analytical level of understanding to the provision of both enhanced design solutions and methodologies to identify system failures; failures that might be fatal. It has been emphasised that while pressure surge failures are generally assumed to involve extreme pressure levels, for example in the failure of pumping mains or the fracture of fuel networks, the Amoy Gardens event illustrated clearly that low amplitude transients capable of depleting appliance trap seals are just a scapable of causing fatalities – a timely reminder of the need for good plumbing design, installation and monitoring at a time when rapidly increasing urbanisation is placing extraordinary strains on the installed infrastructure in developing cities worldwide.

The introduction of Active Control offers a whole new range of design opportunities to the engineer, and the demonstration of the first use of this methodology at the O_2 Dome provides a confidence level in the further use of this technology. However Active Control is a well-developed approach in the wider field of pressure surge analysis and its application here may be seen as a measure of the progress that drainage design has made in providing engineered solutions rather than mere code compliance, where the codes are often based on historic 'rules of thumb', lacking the continuity of scientific underpinning.

The development, in response to the SARS event, of a depleted trap seal identification methodology, by effectively reversing the normal view of transients as phenomena to be avoided or suppressed, offers major public health opportunities. The introduction of such methodology within buildings will provide a means of limiting the possibility of cross-contamination in the future. In addition, persistent trap seal identification will indicate the need for remedial action, possibly drawing on the Active Control possibilities offered by use of combined AAV and PTA solutions.

7 Airflow applications of the Method of Characteristics simulation techniques

The Method of Characteristics solutions applied in earlier chapters to the simulation of low amplitude air pressure transient propagation in building drainage and vent systems has been only one of the unsteady full bore or free surface flow applications of the methodology to solve the St Venant equations of continuity and momentum. The techniques developed are wholly consistent with those utilised in the broader pressure surge or historic waterhammer research areas where the MoC approach is the industry standard. Therefore it is useful to extend the treatment described so far to a wider range of airflow applications that share the same necessity to consider time dependent changes in operating condition.

The fundamental approach that any system may be simulated by separate consideration of the effect of boundary conditions, where only one characteristic is usually available to solve with some user defined boundary equation, and internal conduit nodes, where future flow conditions are predicted by a simultaneous solution of the C^+ and C^- equations, Chapter 3, remains the basis of the simulations presented, including applications to specialised below ground structure drainage provision as well as mechanical venting applications and a brief introduction to the simulation of train motion within tunnels.

7.1 Drainage and vent system simulation for below ground structures

The modelling and operation of building drainage and vent systems has been discussed in previous chapters, however significant differences in the operational regime are present when the form of the building includes a significant below ground component involving the possible collection of system waste at the lowest level and pumping to discharge to the conventional sewer system. In such cases natural ventilation may not be sufficient to prevent odour ingress to the habitable below ground areas and alternative venting may be required. In above ground structures the operation of the drainage network will result in a pressure regime that, while displacing and possibly depleting appliance trap seals, will revert to atmospheric pressure when the discharged flows or applied pressure transients abate. In below ground structures there is often the requirement to collect appliance discharges at the lowest level prior to re-pumping to the sewer connection as well as providing a bilge pump capacity to dispose of below ground water ingress to the structure. The provision of appliance trap and system venting in such circumstances introduces requirements not met in the entrained airflow and shear force driven system operation in above ground structures. One possible solution is to provide a continuous negative pressure regime within the drainage system to ensure the prevention of odour ingress into the habitable space below ground and therefore a continuous discharge of ventilation air at ground level. Clearly the level of this applied suction must be such as to not jeopardise the retention of appliance trap seals. Similarly the treatment of faecal waste discharge from w.c.s must be catered for by the provision of vented collection tanks and pumps whose operation is determined by collection tank level switches.

Figure 7.1 illustrates a typical section of a below ground structure drainage and mechanically assisted vent system as found in mass transit underground stations. The installations found within the London Underground system will be used to demonstrate the application of the modelling techniques based on the MoC solutions introduced previously, while site testing from the Green Park station will also be included, Swaffield and Wright (1998a, 1998b).

In underground structure drainage and vent networks incorporating fan assisted ventilation, air pressure transients may be generated by the following operations:

- appliance discharge flows to the network and conveyed to lowest level collection tanks;
- fluctuations in the free air volume within holding tanks in response to both appliance inflows and pumped discharges to the near surface sewer system;
- mode of operation of the ventilation fan, including start up, speed change or inadvertent power failure;
- surcharging of the sumps or holding tanks, for example following a power failure in level sensing pump control.

In order to apply the MoC simulation techniques already developed it will be necessary to introduce an extended range of boundary conditions to supplement those introduced in Chapter 3 and used elsewhere in this book. In addition it will be necessary to define the normal flow direction and introduce suitable simplifications where necessary to allow the modelling of the networks.

The C$^+$ and C$^-$ equations developed in Chapter 3, and Figure 7.2, remain relevant for the networks considered here and are merely restated below. As previously, the C$^+$ equation links conditions upstream at time t to conditions at the calculation node at time $t+\Delta t$, while the C$^-$ equation links conditions downstream at time t to the calculation node at time $t+\Delta t$. Thus at entry to a pipe section only a C$^-$ characteristic exists while at exit from a pipe section only a C$^+$ characteristic exists. At intermediate nodes both C$^+$ and C$^-$ equations are available.

The C$^+$ characteristics linking RP, Figure 7.2 may be expressed as

$$u_P - u_R + \frac{2}{\gamma - 1}(c_P - c_R) + 4f_R u_R \mid u_R \mid \frac{\Delta t}{2D} = 0 \tag{7.1}$$

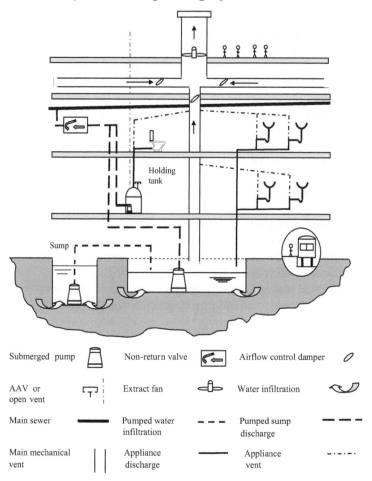

Figure 7.1 Typical below ground drainage system in a mass transit station based on research at Green Park Station, London; note NRV to prevent station inundation in the event of street level flooding

when

$$\frac{dx}{dt} = u_R + c_R \tag{7.2}$$

and the C^- characteristics linking SP may be expressed as

$$u_P - u_S - \frac{2}{\gamma - 1}(c_P - c_S) + 4f_S u_S \mid u_S \mid \frac{\Delta t}{2D} = 0 \tag{7.3}$$

when

$$\frac{dx}{dt} = u_S - c_S \tag{7.4}$$

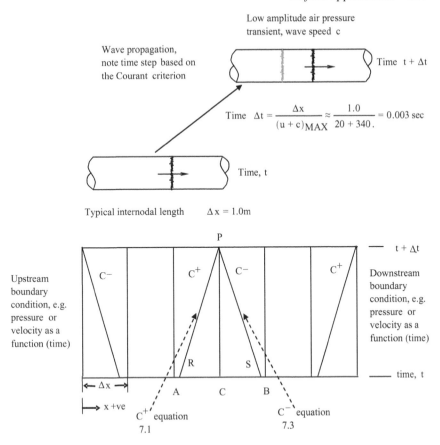

Figure 7.2 Characteristic equations and mode of solution for a pipe section with defined entry and exit boundary conditions

As the variables in the characteristic simulation are wave speed and flow velocity, necessary as pressure and density are interlinked for airflows, it is necessary to determine the appropriate pressure by reference to the expression

$$p_{local} = \left[(\frac{p_0}{\rho_0^\gamma})(\frac{\gamma}{c_{local}^2})^\gamma \right]^{\frac{1}{1-\gamma}} \tag{7.5}$$

From Figure 7.1 the following additional boundary conditions are required:

- mechanical fans, including fan speed change and power failure;
- fan distribution manifold and flow balancing damper installations;
- pump sump operation;
- sealed holding tank operation including any associated local air admittance valves;
- appliance discharge to a pump sump, holding tank or manhole.

The fan boundary condition represents the system to atmosphere interface as it is assumed that the fan discharges direct to atmosphere. The boundary equation at the fan is therefore expressed as the pressure on the system side of the fan, effectively node 1 of pipe 1:

$$p_{t,1,1} = p_{atm} + p_{t,fan} \tag{7.6}$$

which is solved with the entry characteristic, Equation 7.3.

In order to develop the boundary Equation 7.6 it is necessary to introduce the following non-dimensional fan coefficients:

Fan flow coefficient Q/ND^3 (7.7)

Fan pressure coefficient $\Delta p/\rho N^2 D^2$ (7.8)

For a single fan the airflow density and fan diameter may be taken as constant so that the fan delivery flow Q and associated suction pressure Δp at any speed N may be related to the fan reference conditions, defined as Q_{ref}, N_{ref}, Δp_{ref}, from Equations 7.7 and 7.8 as

$$Q = Q_{ref} \frac{N}{N_{ref}} \tag{7.9}$$

and

$$\Delta p = \Delta p_{ref} \left[\frac{N}{N_{ref}} \right]^2 \tag{7.10}$$

The fan characteristic illustrated in Figure 7.3 may be expressed in terms of a polynomial such that

$$\Delta p_{ref} = C_0 + C_1 Q_{ref} + C_2 Q_{ref}^2 + C_3 Q_{ref}^3 \tag{7.11}$$

where the coefficients C_{0-3} are fan specific.

Thus from Equations 7.6, 7.10 and 7.11 the pressure on the system side of the fan may be related to the fan speed and the airflow through the fan if the fan reference conditions and characteristic coefficients are known.

$$p_{t,1,1} = p_{atm} + \left[\left(\frac{N_t}{N_{ref}} \right)^2 \left\{ C_0 + C_1 \left(\frac{N_{ref}}{N_t} \right) Q_t + C_2 \left(\frac{N_{ref}}{N_t} \right)^2 Q_t^2 + C_3 \left(\frac{N_{ref}}{N_t} \right)^3 Q_t^3 \right\} \right] \tag{7.12}$$

By defining fan speed as a function of time within the model it is therefore possible to simulate the effect of fan speed change, including fan start up and shut down, inadvertent power failure or fan speed control based on airborne contamination levels. This model will be returned to in the context of mechanical ventilation systems separately from these drainage network applications. The fan boundary Equation 7.12 is solved with the C^- characteristic, Equation 7.3, to yield time dependent conditions at node 1 of pipe 1 – the system entry.

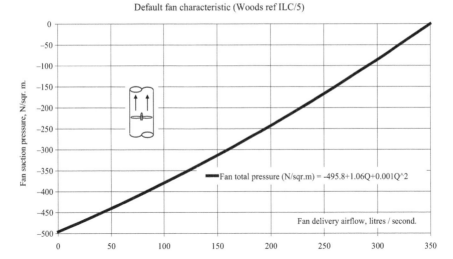

Default fan characteristic (Woods ref ILC/5)

Figure 7.3 Default Woods fan as specified for Green Park underground station, London Underground 1996; Nref = 2590 rpm

Immediately upstream of the fan, any flow balancing to ensure parity through the various ventilation ducts served by the fan will be achieved by the setting of flow dampers on each ventilation duct. For the purposes of this simulation any balancing is achieved by assuming a multi-branch junction with each branch having its individual damper as illustrated schematically in Figure 7.4.

As previously discussed with reference to wet stack entrained airflow, the normal flow direction will be out of the ventilation duct via the fan. The sign convention will therefore be that flow from the network to atmosphere through the fan will be identified as negative flow; this convention therefore matches the identification of the C^+ characteristic available at a duct exit and conversely a C^- characteristic available at a duct entry. The local air pressure will therefore fall towards the fan suction side and then rise to an atmospheric discharge. Figure 7.4 summarises these available equations.

Each damper will have a loss coefficient dependent upon its setting. Thus flow balancing may be achieved by solving the available equations at the junction including the loss terms. This introduces a quadratic solved for each duct entry pressure in turn.

At a junction of n ducts there are $3n$ unknowns, i.e. n pressure, flow velocity and wave speed values and hence there must be $3n$ identifiable equations. The available equations are the C^+ characteristic applicable to the duct terminating at the junction and a C^- characteristic applicable to each of the $n - 1$ ducts originating at the junction – a total of n equations. In addition there is continuity of mass flow through the junction such that

$$\sum_{i=1}^{n} \rho_i Q_i = 0 \qquad\qquad (7.13)$$

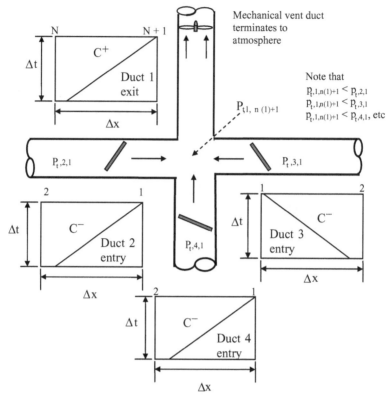

Figure 7.4 Characteristic equations available at the flow balancing junction upstream of the ventilation fan

and for the flow from each of the ducts originating at the junction into the fan duct the following pressure loss expression may be utilised for each combination of ducts 1 to 2 to 1 to n, a total of $n - 1$ expressions:

$$p_{2-n} = p_1 + 0.5\rho K_{(1to2)-(1ton)} Q_{2-n} \left| Q_{2-n} \right| / A_{2-n} \tag{7.14}$$

and n applications of the wave speed – pressure equation 7.5.

Thus a combination of C^+ and C^- characteristics, mass and pressure continuity, including damper imposed pressure loss, the fan boundary equation and Equation 7.5, yields the $3n$ required equations. The absolute value signs used to replace Q^2 with Qabs(Q) ensures that the losses always oppose the local flow. It will be noted that values of p and Q at time $t+\Delta t$ are used as the time step to satisfy the Courant criterion, for duct sections as short as one metre will be small and the error in ignoring an average value across the time increment will be negligible.

Junctions with no applied damper losses may be treated in the same way as junctions in the dry stacks discussed in Chapter 3, effectively the derivation above without the loss terms.

As shown in Figure 7.1, a typical underground station drainage network will feature both open to atmosphere holding tanks and sealed and separately vented holding tanks for faecal and other waste from w.c. appliance discharges. Both forms of tank have to be represented as exit boundary conditions that can incorporate the level variation due to pump operation or flow ingress as well as allowing for the condition where access to the vent ductwork becomes obstructed.

Figure 7.5 illustrates a typical pumped sump tank that will accept water inflows from a system of bilge pumps used to deal with the natural water ingress into the network tunnels – in the London Underground case this is particularly an issue close to the Thames. The sump will also accept appliance discharges excluding w.c. waste. Each sump is connected to the mechanical ventilation system as shown to prevent a build-up of odour in and around the sumps.

The sump boundary condition may be solved with the available exit C^+ characteristic and is based on the continuity of flow expression for the sump, including the mechanical vent, the natural ventilation from the surrounding space and the change in air volume due to water flows to and from the sump.

$$Vol_{sumpair}^{t+\Delta t} = Vol_{sumpair}^{t} + \Delta Vol_{water} + \Delta Vol_{Infiltration} + Q_{vent} \qquad (7.15)$$

The average pressure in the sump across the time step becomes

$$p_{air}^{t+\Delta t/2} = \frac{0.5(Vol_{sumpair}^{t+\Delta t} + Vol_{sumpair}^{t})}{Vol_{Tank} - Vol_{water}^{t+\Delta t/2}} \qquad (7.16)$$

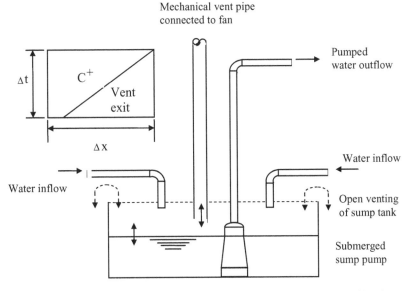

Figure 7.5 Typical sump illustrating the available characteristic and the air infiltration and vent routes

Equation 7.16 may then be solved with the available characteristic Equation 7.1, the pressure–wave speed Equation 7.5 and Equation 7.15, the solution being accomplished via the bisection technique discussed in Chapter 3.

The submerged pump is operated by level sensors, however a failure of this system would result in possible blockage of the mechanical vent duct as the water level rises to submerge its exit. This possibility may also be covered by a boundary condition setting the vent duct airflow to zero, effectively assigning a zero value to airflow velocity, u_p, in Equation 7.1 to yield the boundary condition equation

$$c_P = c_R + \frac{\gamma-1}{2}\left[u_R - 4f_R u_R \mid u_R \mid \frac{\Delta t}{2D}\right] \tag{7.17}$$

The pressure within the sump tank will therefore be determined from the gas laws if it is assumed that the tank free air volume and pressure at time zero are known; the tank pressure is normally assumed to be atmospheric at the start of the simulation. As the sumps are vented locally, variations in sump air pressure will be small and the air infiltration will depend on both the pressure differential to atmosphere and a local entry loss coefficient to represent the degree of sealing provided by the sump covers. A value of 50 was recommended by Wright (1997) for a typical sump cover infiltration loss coefficient. As the degree of fit is loose this loss coefficient will be low and the sump pressure will only be slightly below atmosphere. The design of these sumps reinforces the necessity to provide mechanical ventilation to remove odours. Such sumps do not accept w.c. discharges.

Holding tanks, as illustrated in Figure 7.6, share some of the features of the sumps discussed above. However holding tanks do accept w.c. discharges and are separately vented. Surcharge problems may also exist in the event of pump level sensor failure. In order to prevent the accumulation of foul odour and the depletion of appliance traps during pump operation, a holding tank is always connected to a mechanical vent system and an additional vent pipe, terminated by an AAV, is also often installed to ensure trap seal protection.

The sump boundary Equation 7.15 may be modified to describe a holding tank by removing the infiltration term and extending the vent airflow term to include both the mechanical vent and the dedicated local venting.

$$Vol_{Tankair}^{t+\Delta t} = Vol_{Tankair}^{t} + \Delta Vol_{water} + \sum Q_{vent} \tag{7.18}$$

The tank pressure may then be expressed in a form similar to Equation 7.16 and solved iteratively with the available C^+ characteristic for the exits from the mechanical vent connection and the appliance discharge pipe – which may or may not be actively discharging, and the C^- characteristic for the entry to the local AAV vent pipe.

The free air volume within the holding tank will have a small attenuating effect on any transients propagating within the network, however any change in air volume entering the holding tank will be accompanied by a pressure change so the tank is not the equivalent of a positive transient attenuator (PTA); Chapter 5. Operation of the pumps connected to the system holding tanks will generate

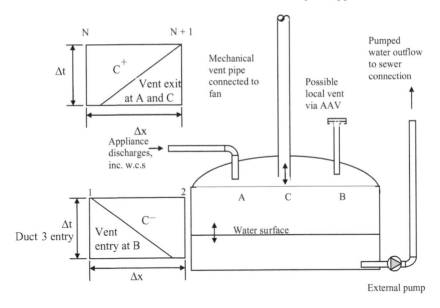

Figure 7.6 Typical holding tank illustrating the available characteristic and the mechanical and natural ventilation routes

changes in free air volume and hence will propagate transients that will travel throughout the network. Similarly any pump or sensor failure may lead to holding tank surcharge and the propagation of severe transients into the network as the mechanical vent airflow is cut off – possibly quite quickly as the water level rises to cover the vent entry.

The air admittance valve installation and operation is identical to that discussed in Chapters 3 to 5. However, in below ground installations material is often a fire concern and the AAVs used may differ from the units commonly applied in building drainage systems. The AAV boundary equations have been developed previously and are not covered separately here.

In most below ground station applications the appliance discharge pipes will be 100 mm diameter, and the analysis on which this treatment is based used earlier work by Swaffield and Campbell (1992a) where the pressure, p, at the discharge from an appliance drain to a sump, holding tank or manhole may be expressed as

$$p = p_{atm} + p_w \tag{7.19}$$

where p_w is the change in air pressure induced by the appliance discharge, shown to be a function of the discharge water flow rate and the diameter of the discharge drain. Campbell (1992) expressed the air pressure at the termination of a discharge pipe as

$$\frac{P_w}{Q_{Water}^2} = B1 + B2\frac{Q_{Air}}{Q_{Water}} + B3\left[\frac{Q_{Air}}{Q_{Water}}\right]^2 + B4\left[\frac{Q_{Air}}{Q_{Water}}\right]^3 \tag{7.20}$$

where the coefficient values were based on laboratory measurements and Q_{Air} is determined from the exit C^+ characteristic, Equation 7.1 for the discharging pipe. Solution is again iterative via the bisection method, Chapter 3.

With reference to the overall system, illustrated by Figure 7.1, it is now possible to identify a number of system operations that will generate low amplitude air pressure transients that may lead to trap seal loss in the connected appliances. These transient generating events may be subdivided into the consequences of normal system operation and inadvertent system failures.

The first category obviously includes all appliance discharges but also includes changes in fan operating point and the selection 'on' of the various pumps responsible for the periodic evacuation of the system sumps and holding tanks.

The second category includes power failure to the fan resulting in an uncontrolled shut down of the vent system and surcharging of any holding tank or sump due to excessive inflows or a failure of the level sensing switching system that results in the mechanical vent becoming closed, generating large suction pressure as the fan driven airflow is cut off.

The discharge of any appliance may be represented by the boundary condition Equation 7.19 when the operation time and duration of the appliance discharge is included in the simulation data and is similar to the approach demonstrated for a conventional drainage and vent system in earlier chapters.

The normal fan operation may be dealt with by including in the data a fan speed vs. time array, Figure 7.7. Interpolation at each simulation time step thus yields the fan speed at that time as a percentage of the fan reference speed and allows the boundary equation to be solved with the appropriate characteristic.

From the available C^- characteristic the airflow through the fan may be expressed as

$$Q_t = A_1\left[u_S + \frac{2}{\gamma-1}(c_P - c_S) - 4f_S u_S \mid u_S \mid \frac{\Delta t}{2D}\right] \tag{7.21}$$

Equation 7.21 may be used to replace the Q_t in Equation 7.22 that equates the pressure appropriate to the wave speed in the available C^- characteristic with that required by the fan characteristic.

$$\left[(\frac{P_0}{\rho_0^\gamma})(\frac{\gamma}{c_P^2})^\gamma\right]^{\frac{1}{1-\gamma}} = P_{atm} + \left[\left(\frac{N_t}{N_{ref}}\right)^2\left\{C_0 + C_1\left(\frac{N_{ref}}{N_t}\right)Q_t + C_2\left(\frac{N_{ref}}{N_t}\right)^2 Q_t^2 + C_3\left(\frac{N_{ref}}{N_t}\right)^3 Q_t^3\right\}\right] \tag{7.22}$$

Equation 7.22 is thus an equation in unknown wave speed and fan airflow rate at time $t+\Delta t$ that may be solved by the bisection method, Chapter 3, to yield the required values of velocity, wave speed and pressure at the entry to the system.

The operation of the sump and holding tanks within the network have been discussed, however in each case there is a possibility of a surcharge event if the vent duct becomes submerged due to excessive inflows of water or due to pump or level sensing failure. Figure 7.8 illustrates this possibility and the appropriate boundary conditions.

Under normal operating conditions the suction provided by the fan will result in an airflow through the ventilated sump or the sealed holding tank, Figures 7.6 and 7.7. In the case of the sealed holding tank the negative pressure in the tank may be sufficient to open the AAV resulting in an airflow, Figure 7.9. The AAV action prevents any trap seals connected to the holding tank from being depleted. In the case of the sumps, airflow is established through the tank ventilation, often provided through metal grid sump covers. If the holding tank or sump becomes surcharged, due to excessive inflows or level sensing or pump failure, then these airflow paths are blocked and air pressure transients are propagated into the main fan assisted vent duct and into the AAV vent connection, the boundary condition to be solved with the available exit C^+ or entry C^- characteristic being

$$u_{t+\Delta t, local} = 0.0 \tag{7.23}$$

Surcharge will generate transients that will propagate throughout the network and may be sufficient to deplete the appliance trap seals connected to the fan assisted vent, Figure 7.1. As discussed previously the magnitude of these transients is expressed by the Joukowsky equation

$$\Delta p = -\rho c_{t+\Delta t, N+1} \Delta u_{t+\Delta t, N+1} \tag{7.24}$$

which for an instantaneous closure, or one completed in less than one relevant pipe period, Chapter 3, equates to approximately 40 mm of water gauge for each 1 m/s destroyed. Note that the transient generated by blocking the inflow to the mechanical vent will be negative while that transmitted into the AAV connection will be positive

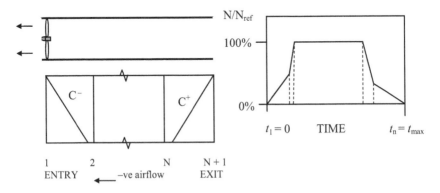

Figure 7.7 Fan boundary condition; note airflow is negative due to choice of sign convention

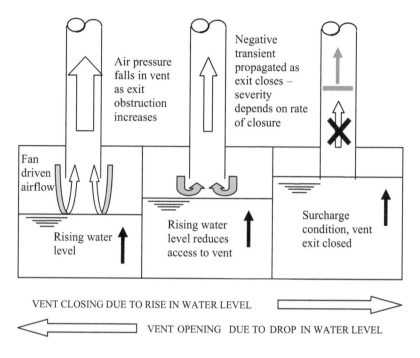

Figure 7.8 Sump or holding tank boundary condition as the rising water level reduces access to the vent exit and then closes it once the exit is submerged; as the water level falls the vent reopens

and will close the AAV, Figure 7.9. It is possible in the case of a holding tank for this sequence of events to result in a trapped positive pressure within the holding tank – a possible hazard in the case of any resulting maintenance activity.

Figures 7.8 and 7.9 indicate the likely process as the tank surcharges and it will be appreciated that closure is unlikely to be instantaneous so that the full Joukowsky pressure surge is not experienced. The simulation is capable of representing this 'slow' closure by means of a loss coefficient K_c to represent the partial closure of the vent access, as shown in Figure 7.8 as the water level rises – applicable to both the mechanical vent and the AAV connection. The boundary condition at the vent–water surface interface may therefore be expressed as

$$P_{t+\Delta t,N+1} = P_{sump} - 0.5\rho K_c u_{t+\Delta t,N+1} \left| u_{t+\Delta t,N+1} \right| \tag{7.25}$$

where the absolute value ensures that the pressure loss opposes motion and the loss coefficient may be approximated by a relationship of the form

$$K_c = K_{Max} \left[\frac{\Delta T_c}{T_c} \right]^n \tag{7.26}$$

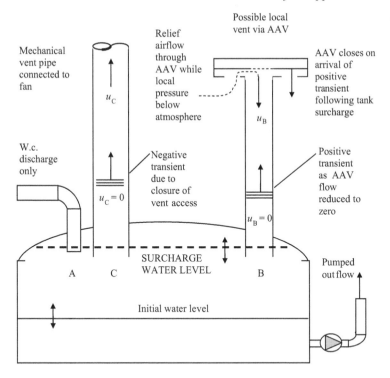

Figure 7.9 Transient propagation as a result of the holding tank becoming surcharged; note that a fall in water level that exposes the entry to the mechanical vent, the AAV vent and any appliance branches generates further transients of reversed sign

where K_{Max} is the maximum loss coefficient achieved during closure (0 for a fully open vent entry and 60000 for a fully obstructed entry, i.e. a figure large enough to ensure a sensibly zero airflow), T_c is the total closure time, ΔT_c is the time into the closure and n is an index to represent the mode of closure.

When the sump or holding tank water level drops, the process described reverses and large transients will again be generated, having the opposite sign to those discussed above. Again it is likely that the opening process will be gradual so the loss coefficient applicable during opening, K_o may be expressed as

$$K_o = K_{Max} \left[1 - \left\{ \frac{\Delta T_o}{T_o} \right\}^n \right] \tag{7.27}$$

where K_{Max} is again the maximum loss coefficient representing a fully obstructed entry and suffix 'o' represents opening values of time.

The choice of the loss index n may be related to the mode of closure or reopening of the vent. A value of $n = 1$ would indicate a linear closure or opening of the vent – an unlikely mode. In the closure mode it is likely that the loss would change slowly at first, becoming rapid over its final stage – suggesting a value of

$n > 1$, while during an opening process the loss would fall rapidly as soon as an airpath appeared and then change slowly, suggesting a value of $n < 1$.

The simulation may be used to investigate all of the transient generating processes discussed, including changes in fan speed and obstruction of the vent system access. Figure 7.10 illustrates a typical underground system station drainage network, loosely based on the system investigated at Green Park, London in 1995 as reported by Swaffield and Wright (1998a and b).

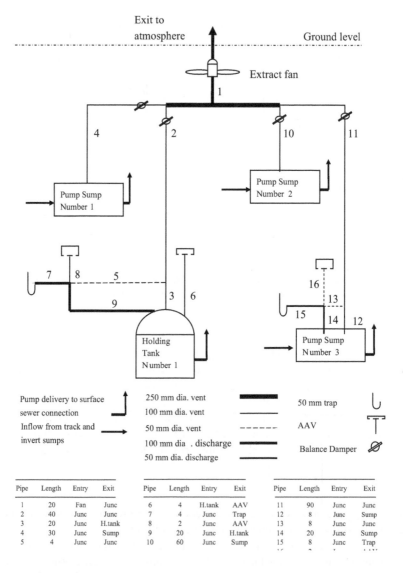

Pipe	Length	Entry	Exit	Pipe	Length	Entry	Exit	Pipe	Length	Entry	Exit
1	20	Fan	Junc	6	4	H.tank	AAV	11	90	Junc	Junc
2	40	Junc	Junc	7	4	Junc	Trap	12	8	Junc	Sump
3	20	Junc	H.tank	8	2	Junc	AAV	13	8	Junc	Junc
4	30	Junc	Sump	9	20	Junc	H.tank	14	20	Junc	Sump
5	4	Junc	Junc	10	60	Junc	Sump	15	8	Junc	Trap

Figure 7.10 Schematic layout of typical underground station drainage network and basis for simulation via model LULVENT

Figure 7.11 presents the simulated airflows within the main vent system as the fan runs up to full speed. The simulation preserves the necessary continuity of flow for the whole network at all fan speeds. It should be noted that in order to discriminate between the main vent pipes, 2, 4, 10 and 11, the flow balancing capability of the model was not used; in practice the dampers provided would have been programmed to an intermediate closure position to ensure equal extraction along each vent duct.

Figure 7.12 confirms the description of the airflow in the pipework close to the holding tank where the ventilation airflow is admitted to the system through the connected AAV as shown.

Figure 7.13 confirms the reduction in line pressures as the fan accelerates to full speed. The AAV opening and subsequent flutter is clearly simulated. As the diaphragm of an AAV is light there is likely to be little damping in practice and the simulation does therefore accurately reflect AAV action.

As expected the simulated pressure in each of the vent ducts falls in stages as the fan runs up to its reference speed of 2590 rpm, Figure 7.3. The AAV opens in response to the reduced pressure in the holding tank and the resulting pressure oscillations simulated in pipe 2 reflect the flutter of the AAV diaphragm at certain airflows – this tends to attenuate as the airflow rises, as shown by Figure 7.14.

The low amplitude air pressure transients propagating throughout the network as a result of fan speed change have the expected effect on the appliance trap seals included in the network. The normal below atmospheric pressure regime within the network implies that the normal steady state condition would be that all appliance traps were slightly depleted relative to the 50 mm norm and may be a reason to introduce deeper traps into such networks. Figure 7.15 illustrates this trap seal loss, although both traps are still operational at these retention levels.

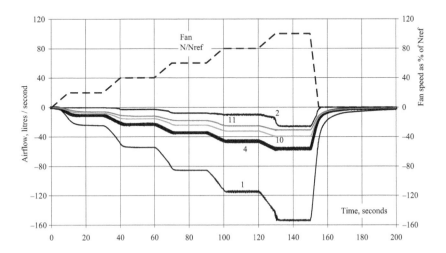

Figure 7.11 Airflow rates predicted within the Figure 7.10 network as the mechanical vent system fan runs up to full speed

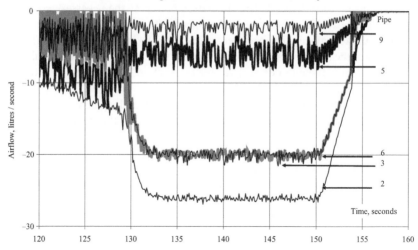

Figure 7.12 Flow continuity preserved around the holding tank, confirming that the pipe 2 extract is drawn through the holding tank AAV

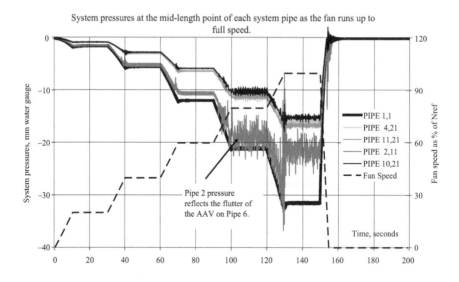

Figure 7.13 Predicted pressures in the main vent ducts as the fan speed is increased by stages, illustrating the response of the pipe 6 AAV

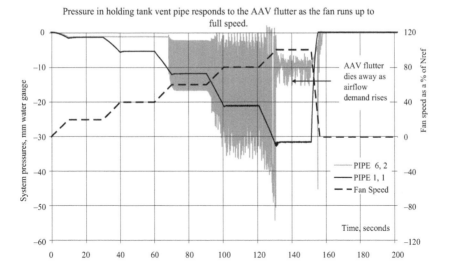

Figure 7.14 AAV flutter as the fan speed increases and the airflow demand through the holding tank AAV on pipe 6 rises

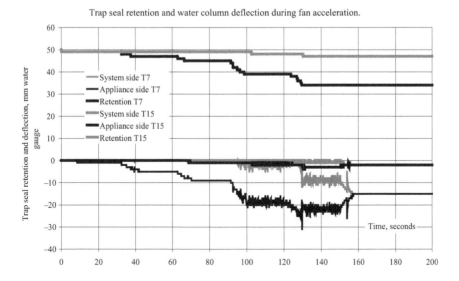

Figure 7.15 Trap seal retention and water column oscillation within the two appliance traps illustrated; note below atmospheric pressure in the main vent ducts depletes both traps

Surcharge of the holding tanks and sumps connected into the network is an obvious source of air pressure transient propagation. Closure of the vent results in large negative transients in the vents feeding directly to the extract fan, while local vents, such as the AAV vent on pipe 6, will demonstrate a positive transient as the airflow through the AAV is destroyed. Figure 7.16 demonstrates this sequence of events; a transient response is seen at the time corresponding to the surcharge of the holding tank, however this does not materially affect the overall airflow in this network as the available inflow route through the appliance vent AAV on pipe 8 provides the required airflow demand. Figure 7.17 illustrates this local network airflow condition.

Consideration of the pressure levels in the network as a result of holding tank surcharge is also possible via the simulation discussed. Figure 7.18 illustrates the Joukowsky type pressure surge generated at the holding tank as it surcharges. The airflow in pipes 2 and 3 is reduced to zero instantaneously and therefore a large negative transient is generated as shown. The presence of the vents joining to form the fan duct provides effective attenuation and only a momentary surge is seen in each of the other main vent pipes. A closer inspection of the pressure levels in the pipes immediately adjacent to the holding tank, Figure 7.19, demonstrates the positive transient propagated into pipe 6 as the tank surcharges; this transient closes the venting AAV and results in the trapped air pressure in pipe 6 already discussed. As soon as the holding tank water level falls, the airflow is re-established through the extract fan and the pressure in pipe 6 again falls to below atmospheric and demonstrates an oscillation due to the flutter of the AAV simulated as the pipe exit condition.

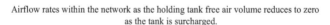

Airflow rates within the network as the holding tank free air volume reduces to zero as the tank is surcharged.

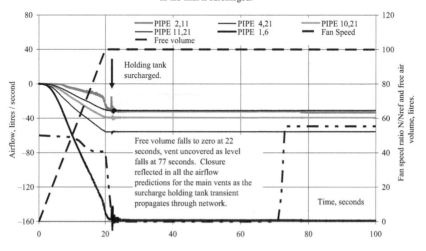

Figure 7.16 Airflow rates throughout the network respond to the closure of the holding tank vent, pipe 6, as the tank is surcharged

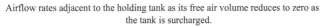

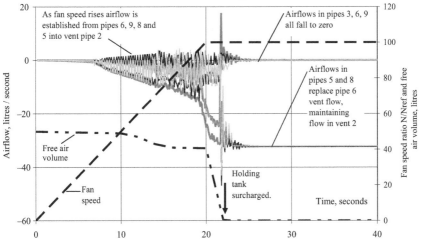

Figure 7.17 Airflow rates adjacent to the holding tank during surcharge closure of the vent, demonstrating the shift in airflow demand to the appliance AAV illustrated in Figure 710

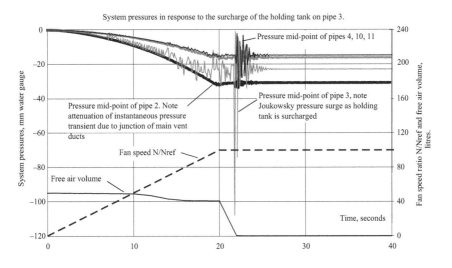

Figure 7.18 Pressure levels at the midpoint of each of the main vent ducts illustrated in Figure 7.10; note the Joukowsky instantaneous pressure surge as the holding tank on pipe 3 is surcharged and the attenuation of this transient by the other vent connections

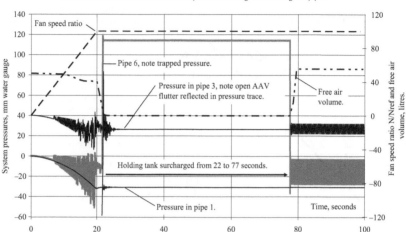

Figure 7.19 Pressure transient propagation as a result of the holding tank surcharge; note trapped pressure in pipe 6 leading to AAV closure and the cessation of the AAV flutter response in pipe 3 for as long as the holding tank is surcharged

Pressure transients propagated in this way have an effect on all the trap seals connected to the network, as already discussed in detail for conventional drainage systems. Figure 7.20 illustrates the additional trap seal loss resulting from holding tank surcharge, compared with the trap seal levels predicted during the fan acceleration, shown in Figure 7.15.

An alternative to an AAV vent local to the holding tank is a long vent to the surface with an open end. This is a problematic solution as the length of the vent mitigates against its effectiveness because the frictional losses will naturally increase with distance. However Figure 7.21 illustrates the difference between the AAV and open vent solution in terms of the local pressures illustrated in Figure 7.19. It will be seen that the trapped air pressure is replaced by a pressure surge oscillation dependent upon the pipe period of the open vent – some 0.3 of a second as the vent is assumed to be 50 m long. The presence of an open end replaces the closed AAV reflection coefficient of +1 with an open end reflection coefficient of –1 so that the transient is allowed to attenuate.

Surcharge of a sump also has similar pressure transient consequences. The main mechanical vent is closed and large negative transients are propagated that may deplete any system trap seals connected to the vent. Figure 7.22 illustrates the effect of such a surcharge on the trap seal retention on pipe 15; it will be seen that the negative Joukowsky transient on vent closure depletes the available trap seal water and allows an inflow of air through the now dry trap. The addition of an AAV to protect trap 15 offsets this total depletion by allowing an inflow to the main vent through the AAV and then pipe 13. Figure 7.22 illustrates this modification

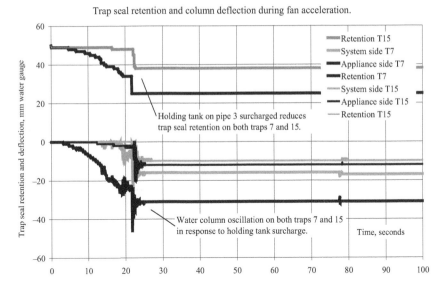

Figure 7.20 Trap seal oscillation and retention as a consequence of the holding tank surcharge; trap 7 is reduced below the acceptable 75% residual level

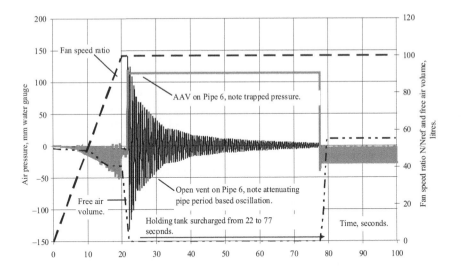

Figure 7.21 Comparison of the pressure transient propagation in the holding tank vent following surcharge with and without an AAV; the open vent is not generally efficient due to the distances to the surface that would be required

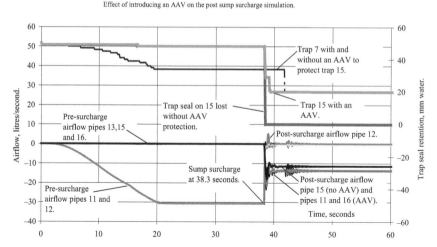

Figure 7.22 Comparison of the trap seal retention adjacent to sump tank 3 with and without an AAV installed to protect the trap on pipe 15; the predicted airflows in the local network also illustrate the effect of the AAV operation

to the airflow paths in the pipe network adjacent to sump 3. It should however be remembered that while AAV installation does offset the problem, choice of AAV material is important in below ground structures subject to stringent fire regulations.

The presentation of an MoC application to simulate the operation of below ground building structure pressure transient propagation has demonstrated that the technique is wholly applicable to this environment provided that the appropriate boundary conditions are applied. The introduction of flow dampers and fan driven mechanical ventilation are modifications to the basic MoC for building drainage and vent systems developed in Chapter 3 and demonstrated in a wide range of drainage applications in Chapters 4 to 6. The opportunities offered to extend the range of the MoC application can therefore be developed; obvious cases would include any mechanical ventilation system while more specialist applications such as VHF medical facilities may also be addressed.

7.2   General application to mechanical ventilation systems, including laboratory fume cupboard extract

It has been stressed that the propagation of low amplitude air pressure transients is a natural consequence of any change in operating condition within a fluid network. The solution of the St Venant equations has been shown to be general and the MoC solution, reliant upon the solution of the system unsteady flow by internal node calculations and simultaneous solution of the available characteristics with suitable system boundary conditions, is also of general application.

The analogy between the shear driven air entrainment in a building drainage system vertical stack and the operation of a mechanical vent is illustrated in Figure 7.23. The main divergence lies in the representation of the driving function. In the mechanical vent the driving function is quite simply the fan. In the annular water flow case the driving function is the shear interaction between the annular water flow and the air core. Effectively this is the equivalent of an infinite number of small fans each contributing in series to the pressure recovery in the dry stack. The sources of pressure loss are simpler to understand as the frictional and minor losses along the drainage system can be seen to be directly analogous to the friction and damper losses in the mechanical vent.

The St Venant equations and their MoC solution are therefore identical in both cases, the boundary conditions alone require consideration. The frictional losses are dealt with using the Colebrook–White equation to determine friction factor and the damper losses dependent upon damper setting, and time if the damper is operated during the simulation, may be modelled using the expressions already defined for the underground structure drainage network. It is assumed that the damper loss coefficient may be defined by

$$K = K_0 + \left(K_{0.5} - K_0\right)\alpha^n \tag{7.28}$$

where K_0 is the fully open loss, $K_{0.5}$ is the 50 per cent open loss, α is the open setting of the damper and n is an index, taken as 8.

Fixed damper or exit/entry grilles may be represented by a simple pressure loss coefficient. Fan operation is modelled as discussed for the below ground structure

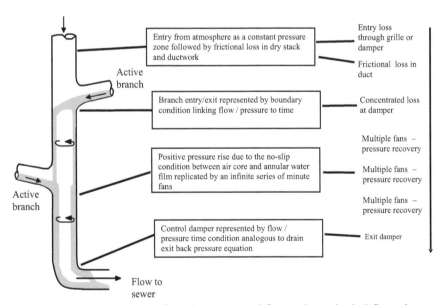

Figure 7.23 Analogy between fan driven system airflow and entrained airflows due to annular water flows

drainage mechanical vent network by relating the fan speed to time and the fan reference speed; Equations 7.12 and 7.22 apply equally to this application.

Figure 7.24 illustrates a mechanical ventilation system incorporating a degree of recirculation. It is assumed that the space remains at constant pressure and that the fans are identical. The system airflows increase as both fans accelerate up to their operating condition. Subsequent movements of the system recirculation damper control the airflow contribution from the room extract and the outside atmosphere, as illustrated. It will be seen that the choice of positive airflow direction results in a positive airflow in supply duct 2 and extract duct 8. Similarly the recirculation flow is taken as positive through duct 7. The time taken for the fan to reach reference speed and the operating times of the dampers are short in this simulation in order to provide a demonstration of the model. However, due to the high wave speed and the relatively short lengths of ducts modelled, slower changes in boundary condition would not result in simulation output much different from that presented, effectively the boundary changes will always be 'slow' in terms of the system pipe periods, as discussed in Chapter 2.

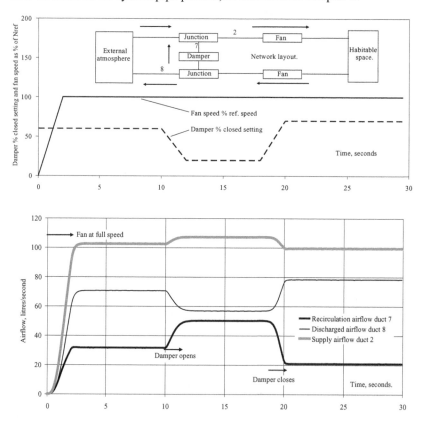

Figure 7.24 Application of the MoC simulation to model recirculation airflow in a mechanical ventilation supply and extract network

While the below ground drainage mechanical vent system simulation was based on the LULVENT model generated as part of a London Underground research programme, the simulation of the recirculation airflows in Figure 7.24 was generated through a freely available simulation, FM4AIR, distributed as part of a general fluid mechanics text, Douglas, Gasiorek and Swaffield (2001).

Operating the recirculation damper has the effect of controlling the percentage mixture of outside and recirculation airflow provided as a supply to the habitable space and could be controlled by gas content monitors in the space and the extract airflow. These control cases could be easily simulated within the FM4AIR model.

The simulation may also be applied to networks of extract ductwork – a typical and common example being laboratory extract from fume cupboards or other zones where controllable and rapid ventilation may be necessary. Figure 7.25 illustrates a typical example of the application of FM4AIR to such an application.

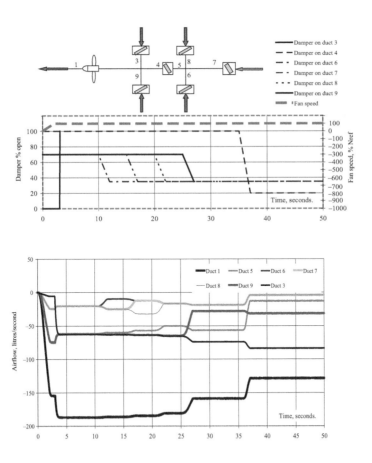

Figure 7.25 Application of the MoC simulation to model the operation of a laboratory extract duct network

An extract fan is positioned at the exit from a network of ventilation ducts, each of which terminates in a controllable damper.

For the purposes of this demonstration it is assumed that each of the network dampers is activated in turn, changes in local extract airflows from dampers 6, 7, 8 and 9 are seen commencing at 5 second intervals from 10 seconds onwards. Note that as the airflow from duct 6 is decreased so the extract from ducts 7 and 8 increases slightly, demonstrating the interlinkage in flow balancing. It should again be noted that the negative sign indicates extract airflow.

Partial closure of the isolation damper 4 at the junction of ducts 4 and 5 reduces the extract from dampers 6, 7 and 8 while the airflow from dampers 3 and 9 increases – as expected.

Thus the MoC solutions and simulations presented directed towards conventional building drainage and vent system analysis have been shown to apply equally to mechanically driven ventilation and below ground drainage vent systems. The solutions to the St Venant equations employed have been identical to those introduced in Chapters 3 and 4 and respect the interrelationship between air pressure and density that results in the necessity to solve the St Venant equations in terms of wave speed, c, and fluid velocity, u, as the dependent variables, pressure being subsequently calculated from the predicted wave speed.

This attention to the pressure–density relationship introduces complications to the MoC solution not found in the general waterhammer or pressure surge literature as normally liquid densities are considered sensibly constant and so the St Venant equations may be solved directly for pressure, p, and fluid velocity u, i.e. directly from Equations 3.2 and 3.4 reproduced below in their general waterhammer form excluding the effects of pipe slope, Table 3.1.

The continuity equation may be expressed as

$$\frac{\partial u}{\partial x} + \frac{1}{A}\left[\frac{\partial A}{\partial t} + u\frac{\partial A}{\partial x}\right] + \frac{1}{\rho}\left[\frac{\partial \rho}{\partial t} + u\frac{\partial \rho}{\partial x}\right] = 0 \tag{7.29}$$

where the second and third terms may be combined, as demonstrated in Chapter 2, by introducing the wave speed in the fluid/conduit combination defined by the fluid bulk modulus K and the conduit geometric and elastic properties, diameter, wall thickness and Young's modulus E, to yield a continuity expression

$$\rho c^2 \frac{\partial u}{\partial x} + \frac{1}{\rho}\left[\frac{\partial \rho}{\partial t} + u\frac{\partial \rho}{\partial x}\right] = 0 \tag{7.30}$$

Similarly the equation of motion for a low amplitude air pressure transient may be recast when the frictional term is represented by the Colebrook–White friction factor as

$$\frac{1}{\rho}\frac{\partial p}{\partial x} + \left(\frac{\partial u}{\partial t} + \frac{\partial u}{\partial x}\frac{dx}{dt}\right) + \frac{4fu|u|}{2D} = 0 \tag{7.31}$$

Equations 7.30 and 7.31 combine, as discussed in Chapter 3, into a pair of total differential equations solvable numerically via the MoC

$$\frac{du}{dt} \pm \frac{1}{\rho c}\frac{dp}{dt} + \frac{4fu|u|}{2D} = 0 \qquad (7.32)$$

provided that the Courant criterion is satisfied, namely

$$\frac{dx}{dt} = u \pm c \qquad (7.33)$$

Wiggert and Martin (2004, 2008) employed this latter formulation, while acknowledging the basis provided by the published LUL research project, Swaffield and Wright (1998a and b), to model extract ductwork systems similar to those considered in the FM4AIR demonstration above, Douglas et al. (2001). The pressure transients observed and predicted were in the same overall range as those encountered in the LUL discussion – transients up to 100 mm water gauge and airflows measured up to 300 litres/second – and the time taken to operate system dampers and run the fans up to speed were also in the same range – i.e. less than 10 seconds. The operational results further confirm the findings of the LUL study presented herein that the MoC solution is wholly applicable to low amplitude air pressure transient propagation. The simplification employed by Wiggert and Martin ignores the interdependence of pressure and density in the treatment of air pressure transients, however the low amplitude allows adequate modelling to be achieved. The methodology is however confined to low amplitude studies whereas the LULVENT formulation, shared with AIRNET, is capable of application under higher pressure conditions, including the propagation of transients within networks with moving boundary conditions, such as train and elevator studies.

7.3 Specialist drainage applications

Within normal hospital drainage network provision there is also the need to provide specialist services to support the treatment of highly contagious diseases. An example would be the specialised Viral Haemorrhagic Fever Units or High Security Infectious Disease Units (HSIDU), several of which exist in the UK. In these facilities it is essential that the usable space is held below atmospheric pressure with the provision of suitable isolation access, Advisory Committee on Dangerous Pathogens (1996). In addition it is essential that any waste is carried away directly to incineration and/or sterilisation equipment and that any venting, both of the treatment space and any effluent venting, which must be included at some point in the network, is provided with high efficiency particulate absorption exhaust filters. Figure 7.26 illustrates the principles but not the detail of such systems and it will be appreciated that the simulation of the transient or quasi-steady flow conditions within such a system and through its fundamental components have been discussed in the treatment of below ground structure drainage networks and the mechanical ventilation networks as represented by the laboratory exhaust system already modelled.

The provision of mechanical space ventilation and the control of the below atmospheric pressure within the space may be determined from a simulation of the

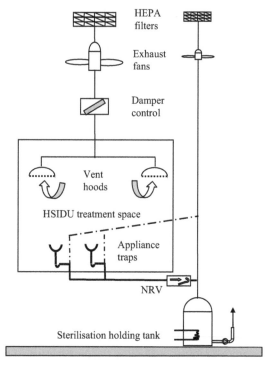

Figure 7.26 Schematic of the principle of operation of an HSIDU drainage and mechanical vented system

mechanical ventilation system. The exhaust ventilation rate is dependent upon the fan control and the damper settings. Failure of any component may be simulated as demonstrated by the effect of isolation damper closure in Figure 7.25. It would be necessary to simulate the resistance of the HEPA filters for both the mechanical ventilation system and the drainage vent system.

Operation of the drainage system involves careful segregation of all patient effluent and its disposal by solidification in absorbent gel and incineration if no effluent holding/treatment tanks are installed. The operation of such systems involves components already simulated in the below ground structure drainage networks. Again it will be necessary to simulate the fan operating conditions as well as the resistance properties of the HEPA filters and any time dependent changes in these values. As the space is held below atmospheric pressure it will be necessary to introduce non-return valves in the event of fan failure. Again the ability to simulate the effects of system failure or surcharge is probably the simulation's greatest advantage. Similarly the simulation allows system modifications to be assessed prior to any reconfiguration of the network or its loading.

The MoC simulation has therefore been demonstrated to have application to a wide range of low amplitude air pressure transient applications provided the system boundary conditions are capable of representation in terms of pressure vs.

flow or flow property vs. time. The MoC simulation techniques will also be shown to apply equally to applications where the required boundary conditions are free to move within the system, such as the simulation of train or elevator transport.

7.4 Applications involving moving internal boundary conditions

The applications of the MoC simulation to low amplitude air pressure transient propagation has assumed that the boundary conditions considered remain fixed in space at the entry and exit from any individual duct or pipe length. While this is appropriate in the cases dealt with so far in the book, there exist a series of applications and potential applications where this would not be the case. Two cases where a moving boundary would be necessary are the prediction of the pressure regime within train tunnels and in high-speed lift shafts. In both these cases the train or the lift car would require to be represented by loss equations linking the longitudinal pressure differential across the vehicle to conditions upstream and downstream and to the velocity of the vehicle.

A further complication, considered later, concerns the assumption of isentropic flow so far justified due to the low amplitude nature of the transients considered. It is recognised that in some cases it is necessary to introduce the energy equation to supplement the continuity and momentum expressions so far considered in order to represent the dependence of wave speed on pressure, density and temperature.

The application of the MoC solution to train motion within tunnels has been a continuing theme of pressure transient research over the past four decades, as exemplified by Fox and Henson (1971) and Henson and Fox (1974a, 1974b), Woodhead, Fox and Vardy (1976), Vardy (1976), William-Louis and Tournier (1998) and Vardy (2008). The introduction of a moving boundary condition requires the addition of the boundary path within the x–t grid normally associated with the MoC solution. Figure 7.27 illustrates the concept for the passage of the vehicle through a grid section. It will be seen that the slope of the vehicle path in the x–t plane may be defined by

$$\frac{dt}{dx} = \frac{1}{V} \tag{7.34}$$

where V is the vehicle mean velocity across that time step.

In order for the simulation to proceed it is necessary to solve the vehicle boundary equation, probably in the form of a pressure differential vs. vehicle speed relationship, with the C^+ and C^- characteristics that exist at that time at location P*, as shown. This requires a knowledge of the base conditions at R* and S* upstream and downstream of the vehicle position one time step earlier. As R* and S* are highly unlikely to fall on node points, for which the base values of u and c are known, it follows that interpolation will be necessary to determine these values. As discussed in Chapter 3, the use of interpolation is questionable as it introduces rounding errors into the simulation as transients arriving at upstream and downstream nodes are taken to affect the interpolated conditions at some

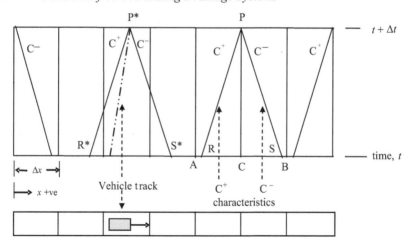

Figure 7.27 MoC x–t grid illustrating the addition of the vehicle track

intermediate point at a time earlier than that at which these transients would have arrived. Linear interpolation is the crudest of the options, however Maxwell Standing (1986) introduced improved interpolation techniques that were shown to yield acceptable simulations in an application to unsteady free surface flows – a challenging case as the surface profile may alter rapidly close to hydraulic jumps. Time line interpolation is also of potential value in this application; the characteristics are extended back through the time distance grid and base points determined for the upstream and downstream nodes from known values several time steps in the past, Goldberg and Wylie (1983), Figure 7.28. Similarly the motion of discrete solids in free surface building drainage under the action of appliance discharges that generate waves that attenuate along the length of the drain requires the introduction of solid path characteristics and the possible interpolation to determine fluid characteristic base values one time step earlier, Swaffield and Galowin (1992).

The methodology to determine base conditions for the C^+ and C^- characteristics is illustrated in Figure 7.29, while Figure 7.30 illustrates the passage of a moving boundary through an x–t grid, with the vehicle moving from rest to achieve a steady velocity and then decelerating.

It will be appreciated that, in common with the interpolation techniques discussed in Chapter 3, it is necessary to know the base conditions at every node at time zero. Normally in this application, as well as the others discussed, the initial start condition is taken as a quiescent condition with pressures set to atmospheric and flow or vehicle velocity set to zero. Other boundary conditions, such as fans, would normally be set to a no-flow condition.

It will be appreciated from Figure 7.30 that particular problems of interpolation may exist at the entry and exit from the overall x–t grid, Figure 7.31. Clearly the characteristics and the vehicle path cannot cross so, in cases where it is not possible to extend the characteristic back to a time at which base values

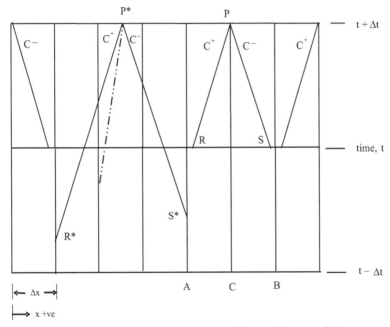

Figure 7.28 Time line interpolation to determine vehicle track base conditions

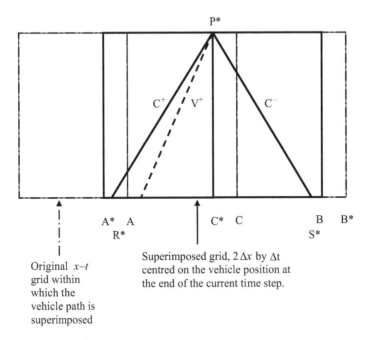

Figure 7.29 Superimposed grid centred on the vehicle position at the end of the current time step to illustrate interpolation requirements

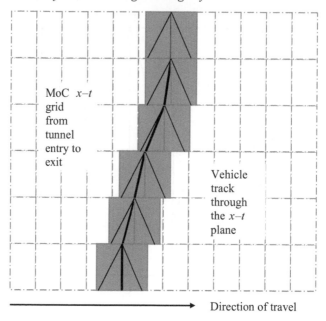

Figure 7.30 Vehicle track superimposed on the system x–t grid to illustrate movement of the vehicle and the shifting interpolation limits at each time step; the interpolation grid either side of the vehicle position moves with the vehicle and is shown greyed on this and later figures

would be known, it is necessary to introduce time line rather than distance based interpolation, despite the fact that this will introduce higher order solutions as the time line interpolation will of necessity include unknown conditions at the end of the time step, Vardy (1976). Figure 7.31 illustrates these issues close to the tunnel exit boundary.

The boundary equations at entry and exit from a tunnel would normally be related to the local ambient pressure, Woodhead et al. (1976). The boundary equation describing the moving train front and rear interface would be based on an application of the energy equation to include the head losses due to air expansion behind the train and coefficient of contraction effects ahead of the train, Henson and Fox (1974a). Woodhead et al. (1976) also considered the effect of a variable resistance exit boundary condition, a water curtain to absorb the incoming transient generated by the approaching train. (In fact a similar curtain at the entry boundary would also be helpful in preventing wave reflections.) The possibility that this would also provide a degree of cleansing was not included in the study. The effect is similar to that exploited in the trap seal retention research discussed in Chapter 6, as the water curtain would present a differing reflection coefficient and could be tuned to reduce any reflection. This concept may have application elsewhere.

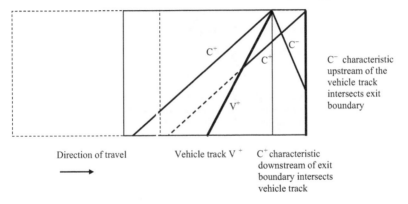

Figure 7.31 Necessity for time line interpolation as the vehicle track approaches a stationary boundary, in this case the tunnel exit

The construction of high-rise buildings and the 'race' to erect the world's tallest building introduces a new application for the techniques discussed herein. Lift or elevator design is moving towards 'super-high-speed' elevators, Bai, Shen and So (2005), where the longitudinal pressure differentials across the elevator cars are of concern due to the blockage factors being generally greater than those encountered in ground transportation studies. In addition the importance of open area ratios, effectively the venting of the elevator shaft has been given consideration. In both these cases the MoC solutions discussed in this book may be of application.

Generally the assumption that the wave speed may be defined by $c^2 = dp/d\rho$ has been shown to be sufficient, allowing the characteristic form of the continuity and momentum equations to be used to predict unsteady flow conditions. However as vehicle Mach numbers rise there is a growing acceptance that the equations of continuity and momentum are not sufficient and the energy equation is added to adequately represent compressible flows where the fluid density depends on temperature and pressure, Vardy (2008). Therefore the three required equations are

$$\frac{\partial \rho}{\partial t} + \frac{\partial (\rho u)}{\partial x} = 0 \tag{7.35}$$

$$\frac{\partial p}{\partial x} + \frac{\partial (\rho u)}{\partial t} + \frac{\partial (\rho u^2)}{\partial x} + \text{shear} = 0 \tag{7.36}$$

$$\frac{\partial \left\{ \rho (c_v T + 0.5u^2) \right\}}{\partial t} + \frac{\partial \left\{ \rho u (c_p T + 0.5u^2) \right\}}{\partial x} = 0 \tag{7.37}$$

The characteristic form of the continuity and momentum Equations 7.35 and 7.36 remain as developed previously and, as the dependent variables are taken as pressure, *p*, and velocity, *u*, appear in a form similar to Equation 7.32, valid

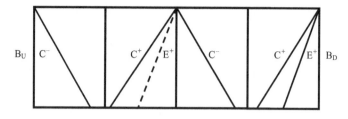

Figure 7.32 MoC solution incorporating the energy equation within an x–t grid

as before along characteristic directions defined by $dx/dt = u \pm c$. The energy Equation 7.37 is expressed as

$$\frac{dp}{dt} - c^2 \frac{d\rho}{dt} - shear = 0 \tag{7.38}$$

valid along a characteristic direction $dx/dt = u$.

The shear terms included in Equations 7.35 to 7.38 again depend upon the wall friction and flow conditions. In order to determine the pressure regime within any duct flow it is now necessary to solve the three available characteristic equations at each internal node. At a downstream exit two characteristics are available but only the C^- equation is present at an upstream entry. Figure 7.32 illustrates these conditions and may be compared with the more traditional characteristics-only approach shown in Figure 7.27. E^+ refers to the energy equation representation and B_U and B_D refer to the upstream and downstream boundary conditions respectively. Note that where two or more ducts join, a sufficient number of characteristics will be available to solve in conjunction with continuity of flow and pressure, as previously discussed in Chapter 4.

Interpolation will again be necessary, however there is a built-in problem. As the slope of the energy characteristic is $1/u$ compared to $1/u+c$ or $1/u-c$ the intersection point will always tend to the centre of the length section and therefore any interpolation errors will be at a maximum. Linear interpolation is to be avoided if possible, however the overall potential error will inevitably depend upon the rate of change of the transient conditions, rapid transients will be severely rounded while 'slow' transients will be less affected. It should also be noted that the application to a simulation including a moving boundary condition may proceed as already discussed, providing the characteristics do not intersect requiring time line interpolation. Overall the addition of the energy equation is important but does raise computational difficulties so its necessity should be assessed for each case.

7.5   Concluding remarks

This chapter has addressed the use of the MoC simulation in a range of applications involving the propagation of low amplitude air pressure transients. The application to below ground structure drainage networks was based on a successful research programme in conjunction with London Underground and demonstrated the application of the simulation to the drainage and vent systems appropriate to below ground structure. Several new boundary conditions were required and in particular the operation of a mechanically assisted vent system was modelled. No issues of application arose and the subsequent agreement between the predicted and measured air pressure regime in the systems was excellent.

The success of the London Underground exercise led to the development of a general ventilation system simulation made available through a mainstream fluid mechanics text and allowed the development of a simulation to address such issues as laboratory fume cupboard extract networks. Again the modelling techniques were shown to be adequate and practical without resort to further simplifications, such as ignoring the interdependence of pressure and density in an airflow subjected to the propagation of low amplitude air pressure transients.

The experience gained through these simulations was also used to show that specialist applications are possible – for example the component parts of the VHF drainage networks installed within High Security Infectious Disease Units (HSIDU) had all been previously developed and tested within the building drainage and vent model AIRNET or its London Underground extension, LULVENT. The application of an MoC simulation to such specialist systems would provide a basis for improved design and staff safety.

Finally the application of the MoC simulations to cases featuring a moving boundary condition was discussed. Moving boundary conditions raise computational challenges, particularly in the vicinity of system boundaries. Earlier work on moving boundaries in free surface flow MoC simulations has shown that care must be taken, particularly in the application of interpolation techniques to identify the base conditions for each characteristic to avoid rounding errors. Examples of train in tunnel simulations were discussed and it was also proposed that similar techniques could be applied to the growing interest in 'super-high-speed' elevators as the number of exceptionally high buildings increases. As Mach number rises with vehicular velocity, the necessity to include the energy equation to supplement the continuity and momentum expressions normally encountered in MoC simulations was also discussed.

Generally the applications covered in this chapter confirm the range of applicability of the method of characteristics as a basis for the simulation of low amplitude air pressure transient propagation.

8   The role of national codes in drainage and vent system design

The complex nature of the unsteady water and entrained airflow conditions within building drainage networks have been discussed in this book. It is important to understand that during the early stages of drainage and vent system development, in the late 1800s, the necessary analytical and experimental monitoring techniques and equipment were not available to assist in the design of safe sanitation systems that would meet the requirement that both odour ingress and cross-contamination were avoided by retention of appliance trap seals. National codes were developed based primarily on 'what was known to work' at the time. While early enlightened researchers such as Putnam (1911) recognised the fluid mechanics principles underlying system operation, in general this insight was not available. Thus, in a Darwinian sense, each 'legislative' zone developed in isolation, with design advances sometimes, but not always, making the transition between zones, for example the US experience of internal systems in the 1920–30 period led to the 'one-pipe system' being adopted in Europe. However the principles of the UK single stack system, introduced in the 1960s and currently also acceptable in Australia, were not recognised in the USA until the 'invention' of the Philadelphia system in the 1990s.

Codes in general continue to be based on 'what works' and are generally developed by committee, inevitably leading to the charge that special interest groups have influence over code development. This may be the basis of an often-expressed view that 'the only way to get a code is to be prepared to compromise'. While national codes have provided, and continue to provide, design guidance that in the main allows system design to fulfil the basic objective of preventing odour ingress and cross-contamination through trap seal retention, the efficiency of any design will vary depending upon the caution that characterised the national code committee. It is said that committees design camels instead of horses, however while both are fit for purpose it might be recognised that one might be more efficient than another in particular circumstances.

Codes should be based on the best available analytical laboratory or site based research information. However, while excellent research facilities existed in the building research centres of most developed countries up to the 1980s, financial cutbacks and the rise of more 'exciting' research agendas has led to a severe lack of well-found research support. At the same time the growing availability of analytical and simulation tools have the potential to provide the community with the option of rethinking codes based on the laws of physics rather than the 'rule of thumb'.

This chapter will highlight some examples where differences exist between national codes that share the same overall objective – these differences must by definition ultimately therefore represent varying degrees of overdesign and code caution. In addition, this chapter will outline cases where the generally accepted content of a code is flawed, and will provide an analytical basis for that opinion.

Codes are essential to continuing public health provision through the design and installation of suitable drainage and vent systems. However it is also necessary to ensure that these codes are based on a correct interpretation of the governing fluid mechanics principles and that they are sufficiently flexible to allow for the inclusion of new concepts and objectives.

In order to initiate a discussion of code development, this chapter will identify the major current codes, their development and the system designations currently employed. The development of techniques to determine likely appliance discharge flows will be discussed, as will the current criteria to determine maximum allowable water throughflows. The essential identification of probable entrained airflows will be reviewed and flaws in current assumptions discussed, with reference to transient theory and the application of fundamental fluid mechanics in the form of the steady flow energy equation. Similarly the sizing of vent stacks will be reassessed in the light of the transient analysis already presented. The criteria determining AAV inflow to systems under suction will be reassessed. Finally, in order to identify differences in code recommendation, a simple, but imaginary, multi-storey system will be discussed in terms of a range of national codes.

8.1   Code development and the need for a unified analytical approach

This chapter will discuss the codes generally used in the UK (and EU), Australia and the USA, a selection that, through acceptance arrangements within various spheres of influence, includes drainage design in many more national jurisdictions. The codes and design guidance documents specifically identified are given in Table 8.1.

As discussed, national codes have developed within their own areas of jurisdiction and represent the degree of caution of the code committee as well as the willingness of the committee, or governing organisation, to take note of current research and development. It is of note that many codes, where reference to base data is included, still reference work of historic rather than current application. A number of choices obviously exist, for example should the maximum allowable waste water downflow in a stack be determined on the basis of historic relationships derived from laboratory work, and hence limited to that set of test conditions, or be based on an allowable pressure regime, thus recognising that the design of the system determines the linkage between entrained airflow, and hence air pressure

Table 8.1 Summary of the codes reviewed within this chapter

Country	Code
UK	Gravity drainage systems inside buildings, Part 2: Sanitary pipework, layout and calculation. BS EN 12056-2:2000
Australia / New Zealand	Australian / New Zealand Standard. AS/NZS 3500.2:2003. Plumbing and drainage, Part 2: Sanitary plumbing and drainage. Standards Australia International Ltd, ISBN 0 7337 5496 1
USA	US Uniform Plumbing UPC 1-2003-1 An American Standard, IAPMO California, 2003
USA	US International Plumbing Code, International Code Council Inc., Illinois, 2003
USA	ASPE DATA Book Volume 2, 2002
	(Note that the USA has a number of codes as well as extensive code descriptive support publications.)

levels, and the applied water flow – these choices are represented by current EU and US codes.

Similarly the choice of allowable trap seal depletion is an important factor; 25 per cent loss is accepted in UK codes, i.e. 37.5 mm always retained in a standard 50 mm trap, while the Australian and New Zealand code accepts a minimum retention of 25 mm.

Venting provision has historically been shown to limit trap seal depletion, however the guidance on vent sizing, particularly in the case of parallel vent stacks, has historically been based on intuition, and a natural aversion to suggesting a vent diameter open to ridicule, rather than analysis. However it may be shown by analysis and simulation that the non-intuitive recommendation that vents should in this particular application have a larger diameter than the wet stack is correct.

Codes are also often reluctant to reflect new concepts. The single stack system, developed in the UK in the 1960s, has not met with universal acceptance – perhaps the similar Philadelphia system, now recommended by ASPE with the caveat that acceptance has to be checked with the local authorities, will prove more acceptable. Similarly there is still hesitation surrounding the use of Active Control techniques, the AAV is still not universally accepted as a means of suppressing negative transients and only the Australian code recognises the P.A.P.A.™ as a means of suppressing positive transients.

Figure 8.1 illustrates the historic development of national codes. It will be appreciated that as systems became more complex, and in the absence of proven analytical or simulation tools in the first half of the 20th century, codes developed by the accretion of 'rules', rather than by radical reinvention to incorporate both new challenges and new analysis methods. The result is that the use of some codes must inevitably result in overdesign. White (2008) compared the UK

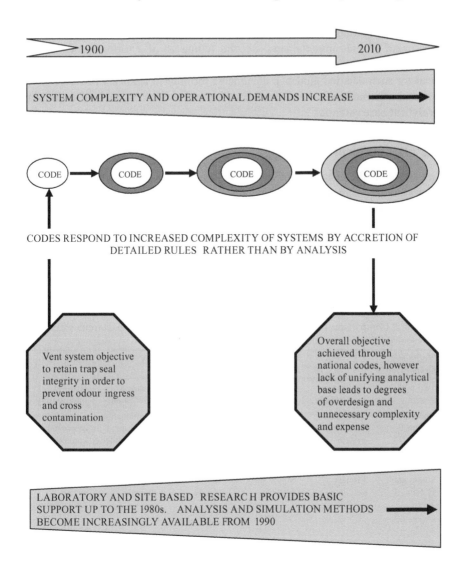

Figure 8.1 Development of national codes during the 20th century led to highly complex and proscriptive codified design guidance

parallel vent sizing code recommendations to those from ASPE, for a 125 mm wet stack, and found that while the BS 5572 UK code (BSI 1994) recommended a 32 mm diameter vent stack, the smallest diameter of vent recommended by the current UK code, EN 12056–2:2000 (BSI 2002) was 50 mm, while the ASPE code recommended a 125 mm diameter vent – clearly there is a degree of caution in the US figures. Neither code yields any real guidance on how to design to a set maximum suction pressure in the stack. While this may have been a problem of limited importance in the past, the new demands for economy coupled with the spread of code imperialism – i.e. the selling on of a national code to new international jurisdictions that may not possess the design capability or the actual need for such detailed guidance documents formulated against a different set of national criteria, make the review of codes and the introduction of a unifying analytical base essential.

8.2   Definition of terminology

In Chapter 1, the historical review of the development of building drainage and vent system design used traditional system descriptors, the terms included the two-pipe, one-pipe, single stack and modified one-pipe systems as well as Active Control to introduce the use of combined AAV and PTA mechanisms to control and suppress transients. The more recent editions of the national codes referred to have introduced alternative terminology and Table 8.2 lists a selection of comparative descriptors.

Table 8.2 Comparison of the 'traditional' system descriptors with current practice as represented by national codes

Traditional terminology	UK (EU)	AS/NZS	UPC	IPC	ASPE
Two-pipe system	Historic interest only	Historic interest only	Historic interest only	Historic interest only	Historic interest only
One-pipe system	Secondary ventilated system	Fully vented modified system	Combination of one-pipe and modified one-pipe as standard	Combination of one-pipe and modified one-pipe as standard	Combination of one-pipe and modified one-pipe as standard
Single stack system	Primary ventilated system	Single stack system	Not included in code	Not included in code	Philadelphia system is fundamentally similar to the single stack system
Modified one-pipe system	Secondary ventilated system	Single stack modified system	Combination of one-pipe and modified one-pipe as standard	Combination of one-pipe and modified one-pipe as standard	Combination of one-pipe and modified one-pipe as standard
Active Control	AAV allowed as part of both primary and secondary ventilated system design up to 10 storeys by BBA cert	AAV and P.A.P.A.™ allowed within code	AAV not mentioned in code but may be allowed locally	AAV allowed for specified conditions	AAV allowed for specified conditions

8.3 Discharge unit or fixture unit methodology for calculating system throughflow of water

The UK code introduces the discharge unit, DU, specified for each appliance type, as a basis for determining the design flows within the system. At any point in the system therefore it is possible to determine the sum of the discharge units representing appliances above the monitoring point, $\sum DU$. The water flow, Q_{water}, is then determined by including a factor, K, to represent the frequency of use of the appliance. This follows closely the description by Wise and Swaffield (2002) referred to in Chapter 1.

Thus in general at a particular point in the system the water flow may be determined as

$$Q_{water} = K\sqrt{\sum DU} \qquad (8.1)$$

The frequency factor is defined for a series of design cases, Table 8.3. Figure 8.2 presents the flow equivalent to any particular DU sum and frequency factor. Any continuous flows may be added to this total.

The Australian / New Zealand code has a similar approach where the water flow is defined by

$$Q_{water} = \sqrt{\sum FU / 6.75} \qquad (8.2)$$

In this case any continuous flow is also allocated a fixture unit (FU) value – e.g. 1.5 litre/second is equivalent to 15 FU.

The US UPC and IPC codes also allocated fixture unit values to each appliance and then provide guidance as to the resulting waste water flow. For example the UPC identifies flows from 0 to 0.47 litres/second as 1 FU, 0.5 to 0.95 as 2 FU, 1.0 to 1.89 as 4 FU and 1.95 to 3.15 as 6 FU. Continuous flows may be added in as 2 FU are taken to represent each 0.064 litres/second of continuous flow.

The individual DU or FU value ascribed to any particular appliance differs from code to code. Appliance development, in particular the concentration on low flush volume w.c.s, requires careful consideration as historic DU or FU values for this appliance type will be invalidated by modern designs and hence drain

Table 8.3 UK code allocation of usage frequency to waste water flow determination from appliance based discharge units

Appliance usage		K
Intermittent use	– domestic	0.5
Frequent use	– hospitals, schools, restaurants, hotels	0.7
Congested use	– public access facilities	1.0
Special use	– dependent on situation, might include laboratories	1.2

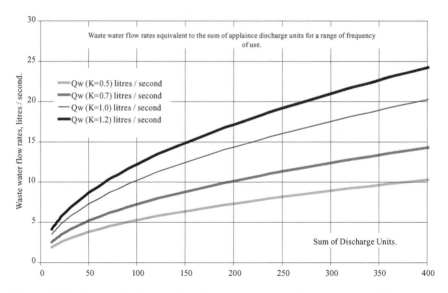

Figure 8.2 Equivalent discharge unit totals and waste water flow for a range of frequency of use cases taken from the UK code

sizing based on these values will overestimate the combined flows and may lead to solid deposition, drain maintenance issues and an impediment to future water conservation efforts.

However the discharge or fixture unit methods do provide a simple translation of the probability based estimation of likely system loading that allows an economic system design to proceed.

8.4   Maximum allowable annular water downflow in vertical stacks

The maximum allowable annular downflow in vertical stacks may be approached from two separate perspectives. Several national codes limit the water flow to that value where the terminal flow conditions would result in a preset percentage of the stack cross-section being occupied by water flow. This approach may be traced back to the work of Wyly at the National Bureau of Standards in the 1960s.

Chapter 1 has developed the relationships linking annular water flow to stack diameter and roughness. Based on experimental work, Wyly and Eaton (1961) suggested that the annular water flow was determined by

$$Q_w = K r_s^{5/3} D^{8/3} \tag{8.3}$$

and suggested maximum flow where the ratio of annular flow area to stack cross-sectional area, r_s = 25 per cent with a value of K for cast iron pipe of 31.9. In SI units, with Q_{water} in m³/s and D in m, this results in values of $1.6D^{8/3}$ for a stack flowing 1/6 full and $3.15D^{8/3}$ for a stack flowing 1/4 full.

Figure 8.3 illustrates these recommendations, naturally limited to the cast iron pipe tested by Wyly and Eaton.

Chapter 1 also introduced values of maximum water flow based on an expression linking water terminal velocity to stack diameter and roughness,

$$V_t = K\left(\frac{Q_w}{D}\right)^{0.4} \tag{8.4}$$

provided that

$$K = \left(\frac{0.2173}{n^2}\right)^{0.3} \tag{8.5}$$

This expression was first developed by Wyly (1964). A typical smooth pipe value of Manning n of 0.007 yields a K value of 12.4, however the more comprehensive Colebrook–White expression suggest a value of K=14.99 for smooth pipes with k=0. Figure 8.4 illustrates the annular flowrate and terminal velocity predictions in this case, while Figure 8.5 presents the vertical distance necessary to attain terminal conditions.

An alternative approach is to limit the water downflow to that value where the entrained airflow results in a maximum suction pressure within the system. The 1994 UK BS 5572 standard set this pressure level at –375 N/m², as discussed by White (2008), however this limit is opaque in the more recent EN 12056–2:2000 code. Despite this, setting a limit based on likely pressure excursions is a more refined approach as it recognises that the pressure level within a system depends upon the resistance to airflow provided by the design itself. Hence, as it will be shown, there is no simple relationship between airflow and applied water

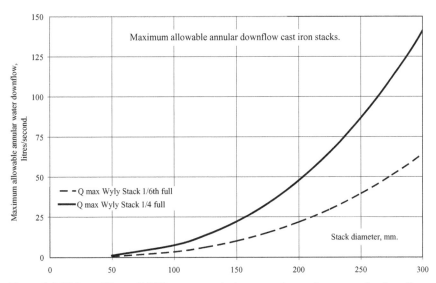

Figure 8.3 Wyly and Eaton (1961) recommendations as to the maximum annular downflow allowable in particular diameters of cast iron vertical stack

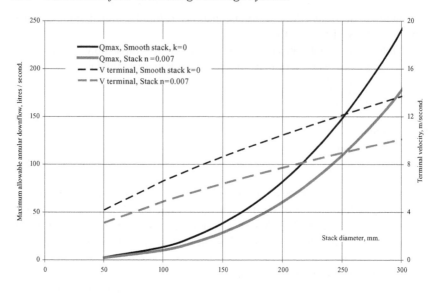

Figure 8.4 Terminal velocity and maximum allowable water downflow in smooth stacks

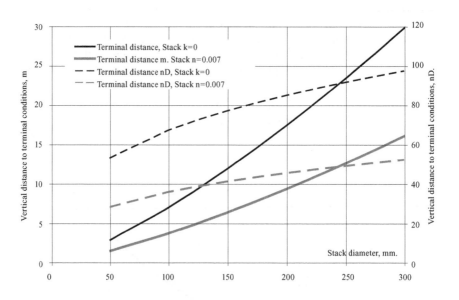

Figure 8.5 Vertical distance required to achieve terminal conditions

downflow. Determining the maximum water flow based on a limiting pressure depression must involve an iterative solution rather than a code 'look-up table'. Designs based on codes that tend to overdesign may not reveal this anomaly, however there is scope here to simplify code recommendations and to economise on system design and materials usage.

Consider a single stack drainage system of 6 or 12 floors. Each floor is represented by a single discharge to the vertical wet stack of 0.25 or 0.5 litres/second so that the total waste water discharge to the stack is 3 or 6 litres/second, uniformly distributed over the 6 or 12 floors. Each floor is 5 m in height and the stack is 150 mm in diameter. The dry stack has an open roof level termination.

Figure 8.6 illustrates a number of important relationships:

1   As the number of floors increases from 6 to 12 with the same total annular water flow, the pressure loss at each junction decreases and the shear forces available to entrain an airflow increase, as the length of the wet stack has increased. The effect is illustrated by comparing the 6 and 12 storey results in Figure 8.6 for a total 3 litres/second water downflow.
2   As the applied water downflow increases from 3 to 6 litres/second for the 12-storey stack, so the junction losses increase, as does the shear force acting on the air core as the annular water surface velocity rises. In this example therefore the entrained airflow may well increase despite the increased junction losses.
3   The radius of curvature of the branch to stack entry junctions on each floor governs the junction loss, supporting the UK code recommendation that

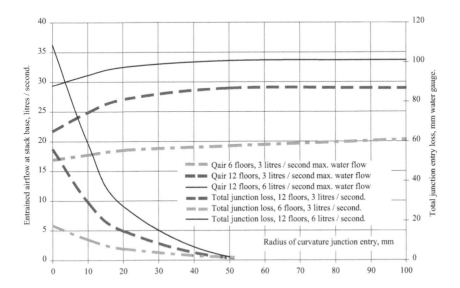

Figure 8.6 Entrained airflow and junction pressure loss as modified by the choice of entry junction radius of curvature and stack height (number of discharging floors)

substantially greater water downflows may be carried by a system with swept junctions.

The mechanism of the junction pressure loss, due to the entrained airflow passing through the cross-stack water film due to the interaction of the stack flow and the discharging branch flow, discussed in Chapter 1, is well understood. The AIRNET simulation used to generate Figure 8.6 utilises the junction loss coefficients identified by O'Sullivan (1974) and re-analysed by Jack (1997).

Thus the detail design of the stack determines the eventual values assigned to the maximum water flow and the degree of airflow entrainment that occurs. There is therefore no unique relationship between entrained airflow and annular water downflow – each case is unique and depends on the energy balance between system losses and shear force action at the water to air interface. This will be discussed in more detail in the following section.

This recognition of the inter-relationship between the system itself and the fluid flows, applied or entrained, was inherently adopted within the UK (1994) code, White (2008), and is at variance with the approach in several US codes where a maximum number of fixture units, representing appliance discharge and frequency, is used as a criterion for maximum applied water flow. The current UK EN 12055–2:2000 is less explicit on this relationship and in comparison with the US and AUS/NZ codes does not give guidance on the appropriate link between vent length and diameter.

8.5   Estimation of entrained airflow

The early research by Wyly and Eaton (1961) at the National Bureau of Standards established empirical expressions for the relationship between annular water downflow and the entrained airflows in building drainage and vent systems. The Wyly expression

$$Q_a = CV_t\left(1 - r_s\right)\pi D^2 / 4 = CK\left(\frac{Q_w}{D}\right)^{2/5}\left(1 - r_s\right)\pi D^2 / 4 \tag{8.6}$$

was developed in Chapter 1 and includes the ratio of mean entrained airflow velocity to terminal mean water velocity, C, that Wyly set to unity. The following expressions were also developed in Chapter 1 to define both maximum allowable water flow, Q_w, and the entrained airflow, Q_a, namely

$$Q_w = K r_s^{5/3} D^{8/3} \tag{8.7}$$

$$Q_a = K_2 r_s^{2/3}(1 - r_s)D^{8/3} \tag{8.8}$$

Figure 8.7 illustrates these expressions for a smooth stack, $k=0$ in the Colebrook–White frictional loss expression, or a Manning's n of 0.007, where the ratio, r_s, of the annular water flow area to the stack cross-sectional area has been set to 25 per cent.

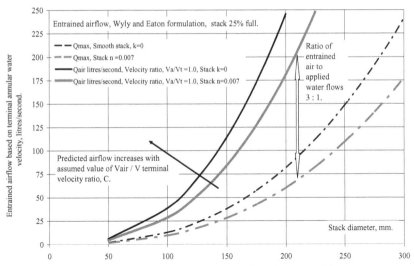

Figure 8.7 Entrained airflow predictions based on Wyly and Eaton (1961)

The ratio $C = V_t/V_w$ represents Wyly's approach to representing the shear forces acting between the annular water film and the entrained air core. However this approach is flawed as it implies a static relationship between the entrained airflow and the annular water downflow, i.e. at $r_s = 0.25$ with $C = 1$, $Q_a:Q_w$ must be 3:1.

Lillywhite and Wise (1969) were the first to challenge this approach by recognising that the stack, from open roof termination to sewer connection, could be represented by an application of the steady flow energy equation which requires that the energy at entry to the system, plus and minus any internal energy transformations, e.g. frictional or local separation losses, must equate to the energy at exit from a control volume constructed around the drainage network. Consider the simple system in Figure 8.8.

If it is accepted that the airflow pressures at entry and exit will be sensibly atmospheric and if the kinetic energy values at entry and exit are negligible, then it follows that the steady flow energy equation reduces to a pressure energy audit between entry and exit that may be defined as

| Pressure recovery in wet stack due to shear forces | = | Σ Frictional losses in dry stack | + | Σ Separation losses at all junctions,entry and exit points | (8.9) |

These terms may be expressed through the standard frictional and separation loss equations as

$$\frac{4\rho f_{wet}L_{wetstack}K(\frac{Q_{air}}{A_{air}} - V_{ter\,min\,al}^{surface})^2}{2D_{wetstack}} = \sum \frac{4\rho f_{dry}L_{drytstack}Q_{air}^2}{2D_{drystack}A_{drystack}^2} + \sum K_{entry,exit,junction} \frac{\rho Q_{air}^2}{A_{stack}^2} \quad (8.10)$$

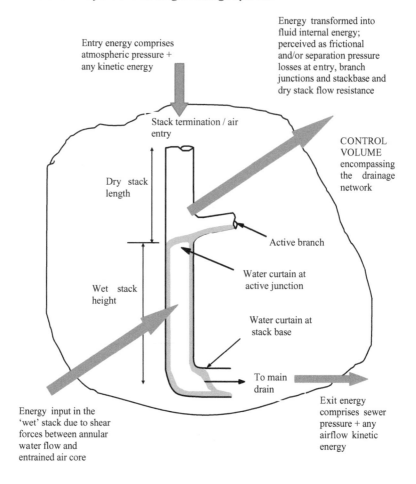

Entry energy comprises atmospheric pressure + any kinetic energy

Energy transformed into fluid internal energy; perceived as frictional and/or separation pressure losses at entry, branch junctions and stackbase and dry stack flow resistance

Stack termination / air entry

CONTROL VOLUME encompassing the drainage network

Dry stack length

Active branch

Water curtain at active junction

Wet stack height

Water curtain at stack base

To main drain

Exit energy comprises sewer pressure + any airflow kinetic energy

Energy input in the 'wet' stack due to shear forces between annular water flow and entrained air core

Figure 8.8 Application of the steady flow energy equation to a control volume constructed around a simple building drainage and vent system

Lillywhite and Wise did not take this approach further as there were no data to infer the value of f_{wet}, an effective 'pseudo-friction factor', essentially negative, that would represent the energy transfer to the entrained airflow. This step, Swaffield and Jack (1998), Jack (2000), was possible following an exhaustive site testing programme where sufficient data at varying flowrates, and degrees of airflow resistance, were collected to allow a 'negative friction factor' to be defined for the wet stack that effectively modelled the pressure recovery that had always been a recognised but not well-explained feature of the pressure regime within building drainage and vent systems.

Thus by rearrangement of Equation 8.10 it follows that the airflow entrained is given by

$$Q_{air}^2 \left[\frac{4\rho f_{wet} L_{wetstack} (\frac{K}{A_{air}^2})}{2D_{wetstack}} - \sum \frac{4\rho f_{dry} L_{drystack}}{2D_{drystack} A_{drystack}^2} - \sum K_{entry,exit,junction} \frac{\rho}{A_{stack}^2} \right]$$

$$+ \frac{4\rho f_{wet} L_{wetstack} K V_{terminal}^{surface} (V_{terminal}^{surface} - 2\frac{Q_{air}}{A_{air}^2})}{2D_{wetstack}} = 0 \qquad (8.11)$$

Thus the entrained airflow is not based on a simple relationship to annular flow percentage and stack diameter but is governed by the overall resistance of the network. The term $K(\frac{Q_{air}}{A_{air}} - V_{terminal}^{surface})^2$ represents a relationship between the surface velocity of the annular film under terminal flow cnditions and the mean velocity of the entrained air core. (This is roughly the equivalent of the standard moving surface example often included in a general fluid mechanics representation of friction loss, particularly in laminar flows where the theoretical integrations are possible, Douglas et al. (2005)). The surface velocity of the annular water flow and the interface velocity of the air have to be equal to satisfy the no-slip requirement, Chapter 3, however the subsequent velocity profile across the air core, and hence the air core mean velocity and volumetric flow, may be variable and take on the appropriate shape to allow varying degrees of air entrainment based on system resistance. Thus restricting the entry at roof level would increase the entry loss and hence reduce the entrained airflow for constant applied water flow and other system parameters, a result amply supported by Swaffield and Jack (1998).

The obvious limiting case for this approach concerns the condition where the upper stack termination becomes blocked. In this case no entrained airflow can enter the stack. Assuming that system traps are sealed, this implies that the air pressure in the upper stack will fall as the shear force continues to act between the annular water film and the air core and the interface air velocity will continue to equate to the surface water velocity. In this case therefore it is possible to postulate an air circulation vertically in the stack so that the air velocity profile reverses with the outer annulus moving downwards with the water flow but the central core moving upwards under the influence of the strong pressure gradient between the stack base, at close to atmospheric pressure, and the upper termination of the stack, which will be well below atmosphere, see Figure 4.22 for a simulation of this condition.

Thus code recommendations based on the entrained airflow predicted by the historic Wyly and Eaton (1961) expressions will be flawed. The simulation techniques discussed in this book allow for the variable air entrainment. While Figure 8.7 confirms that the ratio between air entrained volumetric flow and applied water flows has to be 3:1 under terminal conditions where the value of r_s = 0.25, simulations suggest figures ranging from 8 to 12 times the applied water flow across a spectrum of applied water flows below the limiting value. This result may be corroborated, as shown in Figure 8.9.

If it is assumed that the mean entrained airflow velocity is equal to the applied water flow terminal velocity, then it follows that under the limiting condition set, whether 25 per cent or 16 per cent of stack cross-sectional area, as suggested by Wyly, the airflow will be given by

$$Q_{air} = V_{terminal}\ (1-r_s)A_{stack} \tag{8.12}$$

which by definition will be 3 times the water flow if the annular flow represents 25 per cent of the stack and 5 times when the annular flow absorbs 16.6 per cent of the stack cross-sectional area.

Figure 8.9 illustrates how this ratio varies if the applied annular water downflow is less than the limiting value. It will be seen that while the entrained airflow rises with applied water flow, the ratio of airflow to water flow decreases and effectively covers the range 8–12 observed in the simulations presented in this book. Essentially the expressions are dominated by the area term that results in much higher relative airflow predictions as the annular thickness decreases.

However the 'no-slip' relationship at the air to water interface involves the surface water velocity and the interface air velocity rather than the mean values of either the annular water flow or the air core. The velocity profile in the annular water film becomes important as it dictates the water surface velocity. This will depend upon the roughness of the stack wall; a smooth stack will have a higher terminal and water surface flow velocity. No definitive work is available on the water velocity profile across the annulus. A linear distribution would result in a surface velocity twice the mean or terminal velocity. A more probable parabolic

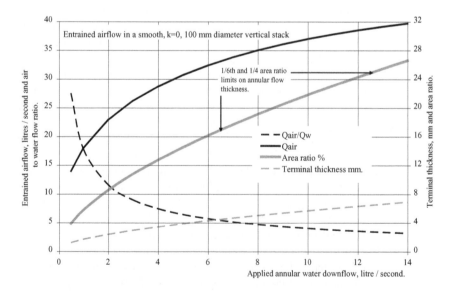

Figure 8.9 Entrained airflow and air to water flow ratio for a 100 mm diameter smooth, k = 0, vertical stack, based on the Wyly and Eaton expression, Equation 8.4

relationship could see the water surface velocity 3 or more times the annular mean, Jack (1997).

No-slip determines the interface air velocity, however the velocity profile across the entrained airflow will vary dependent upon the overall resistance of the network. Figure 8.10 shows that, while the interface velocity of the air and water layers have to be the same, the mean air velocity, and hence the entrained air volumetric flow, may decrease as the air velocity profile 'lags' as shown, i.e. profiles 1 and 2 relate to the orifice openings 1 and 2. This result corroborates the work of Lillywhite and Wise (1969) and Pink (1973b). Figure 8.10 therefore represents the variation in system resistance by a variable orifice at the roof stack entry, an arrangement identical to that used by Jack (1997), and confirms that the concept of a fixed air to water ratio is flawed and cannot be supported as a basis for a code recommendation. Figure 8.11 demonstrates such an effect, based on an AIRNET simulation, confirming the effort shown in Figure 8.6 for variable junction loss.

This discussion, and the simulations presented, therefore confirm that there is no unique relationship between entrained airflow and applied annular water downflow.

8.6   Vent diameter specification

Chapter 5 illustrated the relative performance of the modified one-pipe venting solution compared with an Active Control AAV and P.A.P.A.™ positive transient attenuator installation, introduced to protect a standard single stack system. Table 5.1 confirmed that the most efficient solution was a 200 mm diameter vent in parallel with a 100 mm wet stack, closely followed by an

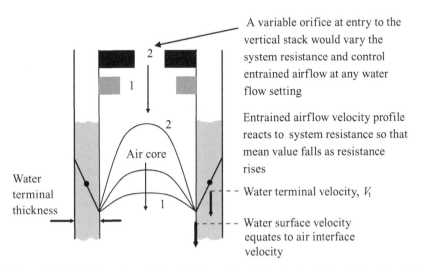

Figure 8.10 Possible water and air velocity profiles in the stack entrained air and applied water flows

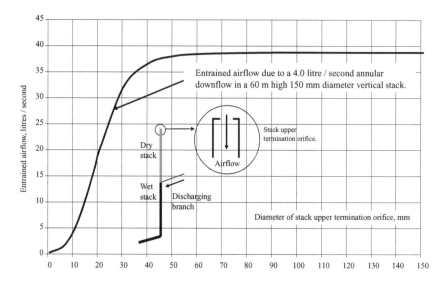

Figure 8.11 Stack entrained airflow as a result of a 4 litres/second annular downflow; note that the entrained airflow is controlled by the assumed resistance at the roof level entry

Active Control solution. Both alternatives were superior to conventional parallel vent systems where the vent was of a smaller diameter than the wet stack. This result is in direct contravention to the consensus recommendation of all the national codes reviewed. Suggesting a dry vent with a greater diameter than the wet stack is counter-intuitive and open to ridicule, however the diversion of the transmitted transient, as shown in Figure 8.12, confirms that with a vent of smaller diameter, commonly 50 per cent of the wet stack value, the reduction in transmitted transient is less than 10 per cent, rising to 66 per cent if the vent to stack area ratio is reversed. This latter solution provides a real reduction in the transmitted transient and is therefore an efficient if costly solution. This solution has little chance of being acceptable, however it may be argued that persisting with current recommendations is merely perpetuating a solution shown to be inefficient when recourse to an Active Control strategy would be equally effective and economic.

The current UK code recommends a 50 mm secondary or parallel vent for wet stacks from 60 to 100 mm diameter, rising to 70, 80 and 100 mm for wet stacks of 125, 150 and 200 mm diameter. The Australian code follows the US example in that the relief vent diameter is increased to allow for the height of the required vent. For a 100 mm diameter wet stack, the diameter of the relief vent depends upon both the stack fixture unit (FU) loading and the length required to reach the upper termination of the vent, as shown in Table 8.4.

The relationship governing the length of the relief vent stack as its diameter increases is the Darcy frictional loss expression,

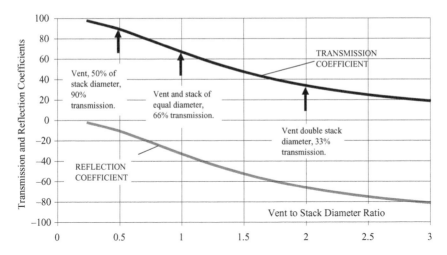

Figure 8.12 Transmission and reflection coefficients at a three-pipe junction depend upon the relative area ratios of the joining pipes and hence vent diameter governs the efficiency of the vent in reducing the transmitted transient

$$\Delta p = \frac{4\,fLQ^2}{2DA^2} = \frac{64\,fLQ^2}{2\pi^2 D^5} \tag{8.13}$$

where the pressure loss due to friction, Δp, depends on the friction factor f, pipe length L, flowrate Q, and the pipe diameter D. Thus for the same pressure drop and flowrate, the length of the pipe is dependent on D^5. (This is of course not strictly accurate as the value of the friction factor depends on both the pipe surface relative roughness, k/D and the flow Reynolds number, $\rho QD / A\mu$, however this effect may be neglected in this application.) Thus the length of the vent is recommended to increase from 9 m to 280 m if the vent diameter is doubled, as (D ratio $= 2)^5 = 32$, and hence the new vent length is actually 288 m.

In terms of increased fixture unit loading, the flow is proportional to the square root of total fixture units, so if the loading is doubled from 150 to 300 then the flowrate increases by 1.4 and the length of the relief vent should decrease from 288 to 206, close enough to the tabulated value of 216 to confirm the rationale.

The US codes use an identical *rationale* to determine vent lengths.

Chapter 4 also include a series of simulations designed to demonstrate the effectiveness of offset bypass or Active Control solutions. Generally offsets should be avoided as they effectively introduce the problems associated with surcharging at the stack base to multiple locations throughout the network. The ASPE venting guide effectively provides the solutions investigated in Chapter 4, with the exception of the combined AAV / PTA case considered. In common with the recommendations on parallel vent sizing, the bypass recommendations

Table 8.4 Examples of the influence of maximum discharge flow and stack height on the chosen diameter for a relief vent, AS/NZS Standard

Stack diameter, mm	Fixture unit total	Maximum height of the relief vent, m			
		Vent stack diameter, mm			
		50 mm	65 mm	80 mm	100 mm
100	150	9	25	70	280
100	300	8	22	60	216
100	500	6	19	50	197

generally refer to a smaller diameter bypass vent or connection into a smaller diameter existing parallel vent. Chapter 4 demonstrated that again an increased vent bypass diameter would be appropriate. The application of Active Control to mitigate the effects of offset surcharge was also demonstrated; one advantage of this solution is that it can be introduced as a remedial action without major disruption to the operating building if found necessary.

8.7   Air admittance valve recommendations

The air admittance valve, discussed in Chapter 5, was introduced in the 1980s to mitigate trap seal loss due to negative air pressure transient propagation. As a surge control device it belongs to the same family as the inwards relief valve commonly found within pressure surge practice, however it has the added requirement that it should seal totally under positive pressure to prevent cross-contamination of habitable space. Doubts as to the self-sealing ability of AAVs has hindered their acceptance in some national codes.

Reference to the UK, AS/NZS and US codes reveals that there are two common requirements in addition to the self-sealing requirement. AAVs are expected to pass an airflow equal to 8 times the stack annular water flow in the vertical stack and twice the water flow in any branch installation. In addition the US codes specify that an open upper termination must also be included in the design to ensure sewer venting.

Scrutiny of the codes fails to reveal why 8 times the annular water flow is the required norm, however de Cuyper (2009) provided an account of the process that led to this criterion being included in the European code. In the case of vertical stacks, based on Wyly and Eaton's NBS Monograph 31 (1961), a limit of 1/6 of the cross-sectional area was chosen as the European limit, yielding an airflow rate of 5 times the water flow if the mean water and air velocities were taken as equal. However it was considered that the mean airflow velocity would exceed the mean water terminal velocity – possibly based on a recognition that it is the surface velocities that are actually affected by the no-slip rule. A value of 1.5 was assigned to the velocity ratio and hence the airflow rate associated with the maximum water annular downflow increased to 7.5 times the water flow rate, rounded up to 8 times.

The determination of branch AAV demand was based on a similar approach drawn from the work of Van Peetersen (1974) who proposed that as the maximum water flow in a branch, determined from the Chezy free surface steady flow equation and based on a depth of flow equal to 50 per cent of the branch diameter and the branch slope and roughness, could be calculated and if the entrained airflow was assumed to have mean velocity equal to the mean water flow velocity then a ratio of 1.0 would be appropriate for the branch airflow. However, the drafting committee felt that the airflow would be higher as the water flow was neither steady nor uniform and opted for a ratio of 2. In hindsight it is likely that Van Peetersen overestimated the airflow as only the air in contact with the water surface, width D under maximum allowable 50 per cent diameter depth water flow conditions, would experience a positive shear force; the remainder of the airflow perimeter, effectively a half circumference of the branch, or $\pi D/2$, would experience a retarding shear force due to the stationary walls of the conduit. A value closer to 0.5 as a ratio might have been more prudent. In addition the actual airflow entrained in the branch would also depend on the upstream entry resistance to the branch and the pressure gradient between the appliance and the branch to stack termination.

These values are important as they determine the design criteria faced by manufacturers of AAVs and if excessive may lead to untenable design objectives.

Similarly, in the US code requirement that any AAV installation also includes an open stack termination, there does not seem to be any recognition of the effect of mixing AAV and open terminations within one network – clearly the entrained airflow will follow the path of least resistance which will continue to be the open stack termination.

A simple application of AIRNET will clarify these comments. Consider a simple single stack drainage system incorporating a 15 m high wet stack receiving waste flow from a single branch and with a further 30 m of dry stack to an upper open termination. Table 8.5 and Figures 8.13 and 8.14 illustrate the entrained airflow and air pressure regime within the network when:

1   the stack is open at its roof termination;
2   the stack upper termination is formed by an AAV;
3   the stack is open at its upper termination and includes an AAV to protect the appliance trap seal on the discharging branch;
4   the stack upper termination is formed by an AAV and the discharging branch retains an AAV.

The addition of a branch AAV has little effect on the total entrained airflow in the stack due to the normal rules of flow distribution. There is a single pressure at the junction so the accumulated entrained airflow pressure losses along the dry stack must equal those incurred by any entrained airflow drawn through a branch AAV and passing along the branch. Thus

Table 8.5 Entrained airflow distribution in a network with an open upper termination, an AAV upper termination and possible branch AAV introduced to protect an appliance trap seal

Stack and branch terminations	Stack base airflow, litres/ second	Upper termination air inflow as % of total entrained airflow	Branch airflow to stack as % of total entrained airflow
Open stack termination, no branch AAV	19.23	100%	0%
Open stack termination, branch AAV	19.23	Effectively 100%	Effectively zero
AAV stack termination, no branch AAV	6.24	100%	zero
AAV stack termination, branch AAV	10.91	40%	60%

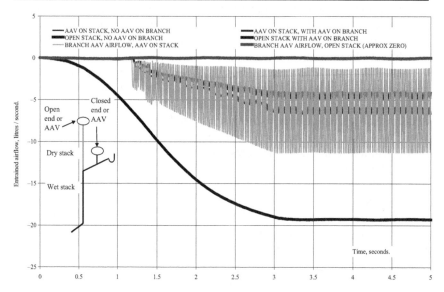

Figure 8.13 Entrained airflow variation dependent upon the choice of stack and branch termination

$$(\Delta p_{friction} + \Delta p_{separation})_{stack} = (\Delta p_{friction} + \Delta p_{separation})_{branch} \tag{8.14}$$

$$\left[\frac{4\rho fL}{2DA^2} + \frac{K\rho}{2A^2}\right]_{stack} Q^2_{stack} = \left[\frac{4\rho fL}{2DA^2} + \frac{K\rho}{2A^2}\right]_{branch} Q^2_{branch} \tag{8.15}$$

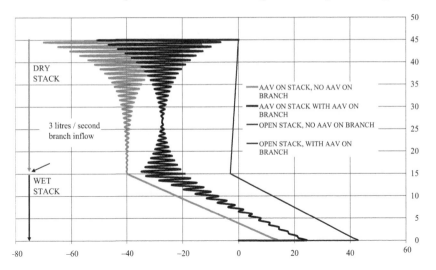

Figure 8.14 Air pressure regime variation dependent upon the choice of stack and branch termination

$$\frac{Q_{stack}}{Q_{branch}} = \sqrt{\frac{\left[\dfrac{4\rho fL}{2DA^2} + \dfrac{K\rho}{2A^2}\right]_{branch}}{\left[\dfrac{4\rho fL}{2DA^2} + \dfrac{K\rho}{2A^2}\right]_{stack}}} \tag{8.16}$$

Therefore the addition of an AAV on the branch while retaining an open upper stack termination would not be expected to vary the stack flow in any major way. Conversely the addition of an AAV at the stack termination will result in the majority of the entrained airflow being drawn through the branch AAV as the pipe lengths are shorter. These expectations are confirmed by Table 8.5 and Figure 8.13.

Figure 8.14 presents the air pressure regime in the vertical stack. It will be seen that the addition of an AAV on the branch has little effect on the air pressure variation up the stack as the entrained airflow entering from the branch is small in comparison with the stack entrained airflow. However the oscillation of the AAV diaphragm is clearly seen as a pressure fluctuation. It will also be seen that the back pressure experienced at the stackbase accurately reflects the changes in total entrained airflow resulting from each design configuration decision.

Similarly the addition of an AAV to the branch and the replacement of the open upper termination by an AAV reduces the airflow in the upper dry stack, however the suction pressures predicted are increased considerably as shown due to the AAV entry loss. Again the upper stack termination AAV demonstrates a diaphragm oscillation that is predicted in the stack air pressure profile. The additional entry loss due to the entrained air entering the stack via the AAV and not an open termination is also seen as a much-increased suction pressure immediately below the roof termination. Again these results conform to the expected variations.

The following observations may be made:

With an open upper termination and no AAV on the branch the pressure regime within the stack is as expected, displaying an initial frictional pressure loss followed by a pressure recovery due to the shear forces in the wet stack. A back pressure is seen at the stackbase as expected. If an AAV is added to the branch then some airflow will enter by this route but the AAV will 'flutter' as it will not be fully open and this will be seen as an oscillation in the stack pressure profile. Otherwise the air pressure regime will be much as for the open upper stack termination.

However, replacing the upper open stack termination with an AAV has a marked effect on the air pressure profile as well as reducing the stack airflow drastically. In the case of a stack AAV and no branch AAV the airflow reduces by over 60 per cent – an effect that would both limit the positive air pressure transients propagated by any stack base surcharge and lower the local stack pressure to which the positive transient is additional – often-overlooked benefits of AAV introduction. The addition of a branch AAV introduces extra airflow, raising the combined stack throughflow by 40 per cent. The frictional loss in the dry stack, at 30 m length, compared with the 2 m long branch will also act to reduce the importance of the upper stack entry.

These predictions of relative AAV inflows have a marked resemblance to empirical data presented by Masayuki and Zhang (2009) who investigated the effect of additional branch mounted AAVs.

8.8   Active Control solutions

Recognising the venting of drainage networks as an example of surge suppression and control allows the designer to access the body of research available in pressure surge theory and practice. Surge suppression by means of relief valves, air chambers and surge tanks is the norm. This realisation led to the development at HWU of the concept of the variable volume containment positive transient attenuator, PTA, which can absorb positive air pressure transients following system surcharges or the arrival of transients as a result of system surcharges or pump operations remote from the building network. The first commercial example of this technology, the P.A.P.A.™ is now available and examples of its use have been included in this book. Taken together with the existing AAV technology this allowed the proposal of an Active Control approach to surge suppression within building drainage and vent systems. While the UK, Australian and some US codes accept the AAV, only the Australian code recognises the advantages of the P.A.P.A.™ technology.

A combination of Active Control and traditional venting also introduces the possibility of sealed building drainage solutions. Originally proposed by HWU as an addition to security sensitive building systems, the first application, involving both AAV and P.A.P.A.™ units, has been to the O$_2$ Dome in Greenwich, a solution appropriate as the roof of the structure could not be penetrated to allow a standard roof open stack termination. It is hoped that future code revisions will follow the innovative Australian example by including this extension to the range of solutions available to the drainage design engineer.

8.9   A simple building drainage and vent system as a code comparator

This chapter has compared the basis for several commonly used national codes. To put the differences between these codes into perspective, a simple imaginary building was considered similar to the public authority housing used in the HWU group's site testing, Jack (1997). A 20-storey building was postulated, each floor comprising residential accommodation and including bathroom and kitchen appliances. Table 8.6 lists the appliances assumed as a basic requirement in such accommodation and the discharge or fixture unit ratings of each under each of the identified national codes. It will be seen that as expected there is a diversity of unit ratings representing the lack of an accepted international standard for allocation of DU or FU values. The equivalence of the UPC and IPC totals is a numerical coincidence due to summation of differences.

The DU and FU totals were then used to generate the recommended design solutions presented in Table 8.7. The differences are immediately apparent, the UK and AUS/NZ codes allowing either a single stack or modified one-pipe solution, and in the AUS/NZ case a further option of a fully vented system, while the US based codes are restricted to a form of modified one-pipe system as already discussed. It is also of note that the UPC code does not appear to recommend, and does not mention, the possibility of partial Active Control through AAV installation.

Table 8.6 Summation of DU or FU appliance ratings for a 20-storey residential block

Appliance	UK	AUS/NZ	UPC	IPC	ASPE
WC (6l)	2.0	6	3	3	6
WHB	0.5	1	1	2	1
Bath	0.8	4	2	2	2
Shower	0.8	2	2	2	2
Sink	0.8	3	2	2	2
Washing machine (6 kg)	0.8	5	2	2	3
Dishwasher	0.8	3	2	2	2
Total per floor	6.5	24	15	15	18
Total for 20 floors	130.0	480	300	300	360

Table 8.7 Comparisons of the recommended design solutions for a 20-storey residential block, based on UK, AUS/NZ and US codes (code comparisons compiled by Dr D.A. Kelly, Drainage Research Group, Heriot Watt University)

Country	UK	Australia/New Zealand	USA	USA	USA
Code	BS EN 12056-2:2000	AZ/NZS 3500.2:2003	ASPE DATA Book	Uniform Plumbing Code	International Plumbing Code
1 floor	6.5 DU	24 FU	18	15	15
20 floors	130 DU	480 FU	360	300	300
	Primary vented	Single stack			
Stack size	125 mm (square or swept entry)	150 mm	100 mm	125 mm	100 mm
Vent stack required	N/A	N/A	Yes	Yes	Yes
Vent stack size	N/A	N/A	100 mm	125 mm	80 mm
Cross-vent size	N/A	N/A	same as ventilating stack	same as ventilating stack	same as ventilating stack
Cross-vent spacing	N/A	N/A	every 10th floor from top	every 5th floor from top	every 10th floor from top
Branch vent	N/A	N/A	40 mm	No specific mention	40 mm
Offset vent	N/A	Not permitted	*	No specific mention	**

* The offset may be vented as two separate soil and waste stacks namely, the stack section below the offset and the stack section above the offset. Offsets may be vented by installing a relief vent as a vertical continuation of the lower section of the stack or as a side vent connected to the lower section between the offset and the next lower fixture or horizontal branch. The upper stack section of the stack shall be provided with a yoke vent. The diameter of the vent shall be not less that the diameter of the main vent or the soil and waste stack, whichever is smaller.

** Above offset – stack should be vented as separate stack with vent stack connected and offset considered as stack base. Below offset – vented by cross-vent between offset and next lowest horizontal branch

Country	UK	Australia/New Zealand	USA ASPE Data Book	USA	USA
	Secondary vented	*Single stack modified*	*Sovent System**		
Stack size	125 mm /100 mm (square / swept entry)	100 mm	100 mm		
Vent stack required?	Yes	Yes	N/A		
Vent stack size	70 mm /50 mm (square / swept entry)	65 mm	N/A		
Cross-vent size	same as ventilating stack	same as ventilating stack	N/A		
Cross-vent spacing	Every floor	Every floor	N/A		
Branch vent	N/A	N/A	N/A		
Offset vent	above and below offset	above and below offset	N/A		
			N/A		
		Fully vented	*requires the inclusion of an aerator at each floor level and an deaerator at the base of the stack. This system has no height limitations.		
Stack size		100 mm			
Vent stack required		Yes			
Vent stack size		100 mm			
Cross vent size		Yes			
Cross vent spacing		not more than 10-floor intervals			
Branch vent		50 mm			
Offset vent		above and below offset			

It is of interest to compare the code approaches to vent stack provision in each case where a parallel vent is recommended, i.e. in the US codes and in both the UK and AUS/NZ codes as an alternative to a single stack system. The UPC code, ASPE guidance and the fully vented AUS/NZ option recommend a parallel vent of equal diameter to the wet stack. In the other cases the vent stack varies from 50 to 80 per cent of the wet stack diameter. None of the codes recommends a vent larger than the wet stack.

All the codes recommend that the cross-vent linking the wet stack to the parallel vent stack should be of the same diameter as the vent stack.

However the frequency of the cross-vent connections varies between codes from every floor (UK secondary venting and AUS/NZ single stack modified) to 5- or 10-floor intervals in the other systems.

The ASPE Data Book (2002) also includes a detailed section presenting the operational advantages of the Sovent system. Originally developed in the 1950s in Switzerland and patented by Fritz Sommer, the Sovent system is utilised in many US installations. Effectively it is a single stack solution that addresses the local pressure losses and oscillatory excursions due to the passage of the discharge water and entrained airflows through successive vertical stack to branch junctions and finally through the stackbase transition to the free surface sewer flow, Schultz (2001). The system is fundamentally similar to the UK single stack system with the important exception that specially developed 'aerator' and 'deaerator' fittings are provided to replace the standard branch to stack junctions and the stack base transition. The internal design of the branch to stack aerator avoids the formation of a water film across the stack that generates localised pressure losses and contributes to the overall pressure regime within the stack. The deaerator fitted at the stack base features a bypass to the downstream horizontal drain to sewer connection that allows a bypass route for the entrained air and hence avoids the positive pressure transient propagation caused by the passage of the entrained air through the stack base water curtain, as fully discussed previously. The stack base transition is also recommended by ASPE (2002) to feature a large radius swept bend or a 45° angled transition, as is also recommended for the UK single stack system. The system is claimed to hold the stack pressure fluctuations within ±25 mm water gauge, Schultz (2001).

ASPE (2002) also makes the claim that the aerator fittings 'slow down' the annular water flow, thus preventing large suction pressure within the stack, a view also put forward by Schultz (2001). While the action of both aerator and deaerator fittings would improve the pressure losses within the stack it is not at all clear how the second claim can be explained.

The modelling of Sovent system operation would be easily achieved if the relevant boundary conditions to define the aerator and deaerator were readily available, however due to the patented nature of the product and minimal UK penetration this extension to the simulation has not as yet been funded.

Clearly, as there are fundamental differences in the recommendations arising from each code for the same physical system and fluid mechanics application, there must be differences in the degree of safety built into each of the code recommendations. Whether such diversity is justified is a point for discussion, however the utilisation of all these codes internationally does raise issues of application – which solution does the designer choose and is there any clear guidance on which to use – the answer is probably 'no'.

8.10 Concluding remarks

National codes are an essential component of the building drainage and vent system design industry. They have developed from initial national responses to the requirement for design guidance within national communities to international documents that now inform drainage design outside their initial jurisdictions. As with any national code the degree of safety built into the codes will vary dependent upon national preferences and perceptions, as well as the historic research base available to any community. While this was not a problem while the codes remained within their own jurisdiction zones, where these cultural preferences and pressures were understood, their application externally to those areas does now pose possible problems where a designer may have several optional directions to develop based on code choice. The opportunity for confusion is clear.

What is also clear from this brief comparison of codes is the lack of a scientific unifying base that may be understood and related back to the laws of physics. Many codes rely on partially understood renditions of historic research that was probably not intended by the researchers to be taken as gospel. Several examples have been identified, perhaps the most noteworthy the clarification that there is no unique relationship linking annular water downflow to entrained airflow, as interpreted by the codes based on experimental work in the 1960s. Rather the relationship should be based on fundamental fluid mechanics, the condition of no-slip and the steady flow energy equation that results in an iterative solution so that the entrained airflow can respond to system resistance.

It is clear that a degree of pragmatism has to be included in code drafting, however such assumptions as are made should be based on a sound understanding of the consequences of a particular choice of simplification. In the case of the assumed relationship between water flow and entrained airflow, the assumptions made in current codes may well impose design criteria in the area of Active Control that are unrealistic and unnecessary.

Overall this review has identified areas where code bodies may wish to look again at their recommendations. The advent of accessible computing simulations, based on fundamental finite difference solutions rather than excessive Computational Fluid Dynamics (CFD) presentations, validated in this area of application as well as in the broader field of pressure surge analysis, does offer fast low-cost comparative advice routes for any study of code modification. It is to be hoped that the drainage and vent system design community can make use of these techniques to 'catch up' the other building services system providers by making available accessible simulation techniques that will remove the reliance on rules of thumb that in many cases cannot be substantiated by analysis based on the accepted underlying unsteady flow fluid mechanics.

However it is recognised that Machiavelli (1513) was prescient when he wrote

> ... there is nothing more difficult to carry out than to initiate a new order of things. The reformer has enemies in all who profit by the old order and only lukewarm defenders in those who would profit by the new order. This arises from fear of their adversaries, who have the law in their favour, and the incredulity of mankind who do not believe in anything new until they have actual experience of it.

Afterword

The *ASPE Data Book* (2002) chapter entitled 'Vents and Venting' opens with the opinion that

> Venting systems are often the least well understood of the basic plumbing design concepts.

This book has attempted to remedy that by offering an alternative approach to the prediction of unsteady entrained airflows within building drainage and ventilation systems, an approach founded on the basic equations of unsteady flow fluid mechanics and numerical modelling techniques recognised as the industry standard for pressure surge analysis and simulation.

It may be readily accepted that venting is the least well understood of the drainage design operations. In order to recognise the root cause of that difficulty it is perhaps useful to consider the process by which fluid flow phenomena are understood. There has to be a logical series of answers to the question 'why?' that lead back to some undeniable truth, such as Newton's second law of motion or the steady flow energy equation. Understanding cannot be achieved if this sequence is interrupted or prematurely curtailed by a 'rule' that states categorically that, for example, the airflow in a stack should be 8 times the applied annular water flow, a contention that cannot be substantiated. Hence understanding is not achieved; instead the practitioner is left with only a set of rules applicable to a particular set of conditions that may not match the conditions under which the design has to operate or do not apply fully to any new or innovative appliance or device to be included in the system.

This book has attempted to correct this lack of understanding by replacing the 'rules' inherent in codified design guidance with arguments based on both steady and unsteady fluid mechanics. By initially drawing the analogy between air pressure transient propagation in drainage networks and the historic Joukowsky approach to pressure surge, or waterhammer, it was possible to introduce the concepts of wave superposition and transmission and reflection at system boundaries, both internal and terminal. These concepts are essential to understanding the pressure regime established within an operating drainage network and the processes that entrain airflow.

The simplified frictionless treatment of transient propagation was shown to be insufficient when dealing with real networks that feature both friction and multiple appliance discharges. It was shown that some form of computer based analysis technique was required to take the argument forward. The Method of Characteristics (MoC) was introduced in the form that revolutionised the treatment of pressure surge in the 1960s – one paper by Lister (1960) launching a thousand research projects over the intervening years. The method was fully described and shown to be applicable to air pressure transient propagation in building drainage networks; the fact that a severe transient might be measured in mm of water gauge rather than in atmospheres was shown to be irrelevant.

Boundary conditions to describe the main features of a building drainage network were then introduced, including physical boundaries such as appliance trap seals, junctions, active or dry, open or closed ends, transient attenuators such as AAVs or PTAs, and boundaries that only arise as a result of the flow conditions, for example stack base or offset surcharge. The centrally important driving functions dependent upon the shear forces between the annular water downflow in the system vertical stacks and the initially quiescent air core were also developed.

The application of the analysis to building drainage was first demonstrated by an extensive range of simulations that dealt with a full range of transient propagating cases, initially in the abstract, in order to validate the applicability of the proposed solution methodology. The results presented are mutually supportive and allow various design cases to be understood – for example why the pressure below a surcharged offset continues to fall – while also demonstrating a transient response based on the appropriate pipe period to the next reflecting boundary condition.

The abstract presentation of the analysis was then further validated by reference to a range of 'real' applications of the model; a forensic treatment of the SARS virus spread; a comparison of active and Passive Control strategies; the development of the P.A.P.A.™ positive transient attenuator into a marketable product and a fundamental constituent of the Active Control strategy; application of Active Control to the O_2 Dome refurbishment; deployment of the analysis and simulation techniques to the specialist field of underground railway station drainage and ventilation provision; demonstration of the applicability of the analysis techniques to specialist health care installations; applications to mechanical venting systems, including laboratory fume cupboards or other ventilation networks, and finally the development of a marketable defective trap seal identification methodology and equipment, a project directly inspired by a need to respond to the dry trap virus spread route for cross-contamination identified as central to the Amoy Gardens epidemic and fatalities – a project in which the HWU team were joined by representatives of code bodies, building services consultants, industry and commercial building operator facilities managers, as well as the UK Department of Health – effectively a coordinated academia–industry institution effort.

Finally this book has attempted to put the outcome of these analysis and simulation proposals in the context of the codified guidance currently available internationally. It is clear that the simulation approach presented could eliminate the need for many of the 'rules' that have become part of current codes by a process of long-term accretion. Replacement of such rules by a logical demonstration of what processes actually occur and why systems respond in the way they do would be a major advantage and would go a considerable way to countering the wholly understandable ASPE assertion that venting is the 'least understood aspect of drainage design'.

The design of building drainage and ventilation networks is therefore at a crossroads. There are increasing demands on the design process; climate change will enforce water conservation; population migration to the mega-cities of the developing world will generate further pressure for accessible and safe systems and system operation; health considerations will dominate design – the SARS epidemic reinvigorated interest in the public health responsibilities of the drainage designer, installer and system operator; the increased complexity of high-rise and prestige building projects internationally will require defensible solutions to new design problems, and finally the need to economise on water, materials and labour costs will bring further pressures to bear. It may be seen that reliance on the gradual accretion of rules within codes, now sold internationally into markets that do not necessarily share the same cultural approach to risk – a form of 'code imperialism' – is not a supportable strategy for the future development of this industry in its widest sense.

The design of building drainage and ventilation networks for complex buildings may be seen as an addition to the expanding remit of the building services engineer and as such should feature at least the same access to predictive analysis as available to designers concerned with the thermal, acoustic or mechanical or natural ventilation systems within buildings. In these allied disciplines the use of predictive simulation models is now commonplace – it is clear that drainage system design lags far behind.

This book has identified the prediction of air pressure levels and entrained airflow within building drainage and vent systems as belonging to a well-understood family of engineering unsteady flow fluid mechanics phenomena that are capable of solution by readily available numerical techniques – in particular the Method of Characteristics (MoC). The book has drawn clear analogies between the analysis of pressure surge in a whole range of fluid networks, from large-scale civil engineering projects concerned with water or oil movement to specialist applications such as in-flight refuelling or fire-fighting systems. In some ways it is possible to extend the analogy as it may be seen that the discipline of building drainage design is at a point analogous to pressure surge analysis in the early 1960s when Streeter and Fox first demonstrated the power of the MoC to simulate and provide an analysis tool to solve pressure surge problems across the whole spectrum of this branch of fluid engineering. Streeter and Fox's work was followed by an international effort to apply these techniques and to develop specialist boundary equations that would expand the field of application to such phenomena as gas release, column separation and pump and turbine operation and failure conditions. The early programs developed by a whole range of researchers were not user friendly packages but research tools in their own right.

This book has detailed the development of transient airflow simulation and prediction within building drainage and ventilation systems using the MoC. The book has included several hundred applications of the model to generate the test cases demonstrated. Over its development since 1989 AIRNET has been used to analyse transient phenomena – for example by providing a forensic validation of the role of air pressure transients in the Amoy Gardens SARS virus spread; as a fundamental constituent of numerous research programmes at HWU; as a basic support tool in the development of new design techniques – such as the below ground building or sealed building system design proposals and the Active Control strategy, first used on the O_2 Dome, that is comparable in efficiency to existing fixed in place vent solutions; as the basis for identifying new transient control and suppression devices – for example the development of the positive transient attenuator (PTA) concept, now embodied in the P.A.P.A.™ device available to system designers as a constituent of new designs or as a remedial solution in the event of system failure caused by positive air pressure transient propagation, and most recently as a central constituent in the ongoing development of the dry trap identification project.

White (2008) was of the opinion that

AIRNET is not user friendly … it is a research tool

That is undeniably true. What is required now is a renewed research effort, supported by industry, government funding agencies and institutions, that will lead to a multiplicity of AIRNET-like simulations as happened in the 1970s in pressure surge and established the methodology as the industry standard. Reference to current ongoing research internationally indicates clearly that there are groups that could take such work forward; groups in Taiwan, Japan, Hong Kong and Brazil have the capability to develop such solutions and have the base knowledge and data to generate their own versions of the central driving functions included in AIRNET that are central to the simulation of the pressure recovery phase of multiple branch to stack system operation.

The research programme at HWU over the past 20 years has laid the foundations for a new approach to drainage and ventilation system design, an approach that will bring the analysis tools available to the designer to a state comparable in efficacy to those enjoyed by the other building services disciplines. There is therefore a challenge to be met by building services and public health academics and researchers, by industry, by government funding agencies and by the code bodies to rethink the approach to building drainage and ventilation system design and to replace rules of thumb and compromise with design techniques firmly based on the science of unsteady flow fluid mechanics.

References

Abraham T. (2005) Twenty-first Century Plague: The Story of SARS, Hong Kong University Press.

Abreu J.M. and Almeida A.B. de (2000) 'Pressure transient dissipative effects: a contribution for their computational prediction', 8th International Conference on Pressure Surges, BHR Group, The Hague, Netherlands, 12–14 April 2000, pp.499–517.

Advisory Committee on Dangerous Pathogens (1996) 'Management and control of viral haemorrhagic fevers', The Stationery Office, London, December.

Allievi L. (1904) 'Notes I–IV translated as Theory of Waterhammer', Halmos E.E. Ricardo-Garoni, Rome, 1925.

ASPE (2002) 'Data Book, Chapter 3 Vents and Venting' Vol 2, Chapter 3, pp.35–65.

Babbit H.E. (1935) 'Tests on the hydraulics and pneumatics of house plumbing', Engineering Experimental Station Bulletin, No 143.

Bai H., Shen G. and So A. (2005) 'Experimental based study of the aerodynamics of high speed elevators', BSER&T, Vol 26, No 2, pp.129–144.

Ballanco J.A. (1998) 'Investigation and analysis of violently fracturing water closets',Technical Proceedings ASPE 1998 Convention, Indianapolis, October 1998, pp.109–150.

Beattie R. (2007) 'Derivation of an empirical frequency dependent friction factor for transient response analysis of water trap seals in building drainage systems', MSc thesis, Heriot Watt University, School of the Built Environment.

Bergeron L. (1932) 'Variations in flow in water conduits', Comptes renders des travaux de la Soc. Hyd. De France, Paris.

Bergeron L. (1961) Waterhammer in Hydraulics and Wave Surges in Electricity, John Wiley, New York.

BHRA (now BHR Group). International Conferences on Pressure Surge, 1972, 1976, 1980, 1986, 1989, 1992, 2000, 2004, 2008. Cranfield, UK.

Billington N.S. and Roberts B.M. (1982) Building Services Engineering: A Review of its Development, Pergamon Press, Oxford.

Boldy A.P. (1976) 'Waterhammer analysis in hydroelectric pumped storage installations', 2nd Int. Conf., Pressure Surges, BHRA, London.

Boldy A.P. and Logan C. (2008) 'Surge chamber issues associated with the design of the Braamhoek hydroelectric pumped storage scheme', 10th Int. Conf. Pressure Surges, BHR, Edinburgh, May 14–16.

Bridge S. and Swaffield J.A. (1983) 'Applicability of the Colebrook–White formula to represent frictional losses in partially filled unsteady pipeflow', Journal of Research, NBS, Vol 88 No 6, Nov–Dec, pp.389–393.

Brock R.R. (1970) 'Periodic permanent roll waves in open channels', J. Hyd. Div. ASCE, Vol 11 No 10, pp.2269–2277.

Brunone B. and Ferrante M. (2004) 'Pressure waves as a tool for leak detection in closed conduits', Urban Water Journal Vol 1, No. 2, pp.145–155.

BSI (1994) 'Code of practice for sanitary pipework', BS 5572, BSI, London.

BSI (2002) 'Gravity drainage systems inside buildings, Part 2: Sanitary pipework, layout and calculation', BS EN 12056-2:2000, BSI, London.

Buckingham E. (1914) 'Model experiments and the form of empirical equations'. Phys. Rev., 2, 345.

Building (Planning) Regulations CAP123I (2000) 'Standards of sanitary fitments, plumbing, drainage works and latrines', Hong Kong SAR Government.

Burberry P. and Griffiths T.J. (1962) 'Service engineering: hydraulics', Architects Journal, 21 November pp.1185–1191.

Butler D. (1991) 'A small scale study of waste water discharges from domestic appliances', Journal Institution of Water and Environmental Management, Vol 5, No. 2,:178–185.

Butler D. (1993) 'The influence of dwelling occupancy and day of the week on domestic appliance wastewater discharges', Building and Environment, Vol 28, No. 1,:73–79.

Butler D. and Davies J.W. (2004) Urban Drainage, 2nd edn, Spon Press.

Campbell D.P. (1992) 'Mathematical modelling of air pressure transients in building drainage and vent systems', PhD thesis, Heriot Watt University.

Campbell D.P. and MacLeod K.D. (1999) 'Investigation of the causative factors of air flow entrainment in building drainage-waste-ventilation systems', Building Service Engineering Research and Technology, Vol 21, Jan, pp.35–41.

Campbell D.P. and MacLeod K.D. (2000) 'Detergency in soil and vent systems', Building Service Engineering Research and Technology, Vol 21, No 1, pp.35–41.

Carstens M.R. and Roller J.E. (1959) 'Boundary-shear stress in unsteady turbulent pipe flow' Journal of the Hydraulics Division ASCE, Vol 85 HY2, pp.67–81.

Chadwick E. (1842) 'Report into the sanitary conditions of the labouring population of Great Britain', London, HMSO.

Chakrabarti S.P. (1986) 'Studies on the development of economical drainage systems for multistory buildings', National Bureau of Standards Washington DC, Amerind Publishing, New Delhi, TT 38-04-000.

Chan D.W.T., Law L.K.C. and Chan E.H.W. (2005) 'Investigation of airflow from drainage stack through empty floor drain trap', ASME/HKIE/IMechE Joint Symposium: Proc. 2nd Symposium on Advanced Technology for Health Care and Hygiene Control, May 27, Hong Kong, pp.65–68.

Cheng C.L., Liao W.J., He K.C. and Lin J.L. (2009) 'Empirical study on terminal velocity of drainage stack', CIBW62 Conference, Water Supply and Drainage for Buildings, Dusseldorf, September 7–9.

Cheng C.L., Yen C.J., Wong L.T., and Ho K.C. (2008) 'An evaluation tool of infection risk analysis for drainage systems in high-rise residential buildings', Building Service Engineering Research and Technology, Aug, Vol 29, pp. 233–248.

Chow V.T. (1959) Open-channel Hydraulics, McGraw-Hill.

Cook G.C. (2001) 'Construction of London's Victorian sewers: the vital role of Joseph Bazalgette', Postgrad Med J 2001;77:802 (December), Wellcome Trust Centre for the History of Medicine, London.

Cummings S., McDougall J.A. and Swaffield J.A. (2007) 'Hydraulic assessment of non-circular section drains within a building drainage network', BR and I Vol 35, No. 3, pp.316–328.

Daily J.W., Hankey W.L., Olive R.W. and Jordan J.M. (1956) 'Resistance coefficients for accelerated and decelerated flows through smooth tubes and orifices', ASME, July, pp.1071–1077.

Dawson F.M. and Kalinske A.A. (1937) 'Hydraulics and pneumatics of plumbing systems', University of Iowa, Studies Bulletin, No. 10.

De Cuyper K. (1993) 'Towards a standardised code of practice for drainage systems inside European buildings', CIBW62 Conference, Water Supply and Drainage for Buildings, Porto, pp.2–13, September.

De Cuyper K. (2009) Personal communication: Required AAV throughflows.

Douglas J.F., Gasiorek J.M. and Swaffield J.A. (2001) Fluid Mechanics, 4th edn, Pearsons.

Douglas J.F., Gasiorek J.M., Swaffield J.A. and Jack L.B. (2005) Fluid Mechanics, 5th edn, Pearsons.

Doyle T.J. and Swaffield (1972) 'Evaluation of the method of characteristics applied to a pressure transient analysis of the B.A.C. / S.N.A.I.S. Concorde refuelling system', Procs. I. Mech E. Vol 186, 40/72, pp.509–518.

Eagle (1951) 'Dan Dare', May 25, Vol 2, No 7.

Edwards K. and Martin L. (1995) 'A methodology for surveying domestic water consumption', Jour. CIWEM, 9 Oct., pp.177–488.

Enever K.J. (1972) 'Surge pressures in a gas–liquid mixture with a low gas content', 1st Int. Conf. Pressure Surges, Canterbury, England, September 6–8, BHRA Paper C1.

Ezekial F.D. and Paynter H.M. (1957) 'Computer representation of engineering systems involving fluid transients', Trans. ASME, Vol 79, Paper No 56-A-120, pp.1840–1850.

Fernandes V.M.C. and Gonçalves O.M. (2006) 'Limits for use of vent elements in building drainage systems considering the risks of infection spread by means of water-seal behaviour and integrity: the case of Brazilian systems', Building Service Engineering, May 2006,Vol 27, pp. 103–117.

Filsell S. (2006) 'Investigations of positive air pressure transients in building drainage ventilation systems', PhD thesis, Heriot Watt University.

Finer S.E. (1952) The Life and Times of Sir Edwin Chadwick, London: Routledge/ Thoemmes Press, 1997 (orig. London: Methuen, 1952).

Fox J.A. (1968) 'The use of digital computers in the solution of waterhammer problems', Proc. ICE, Vol 39.

Fox J.A. (1989) Transient Flow in Pipes, Open Channels and Sewers, Ellis Horwood, Chichester.

Fox J.A. and Henson D.A. (1971) 'The prediction of the magnitude of pressure transients generated by a train entering a single tunnel', Proc. ICE 49, May, pp.53–69.

Galowin L.S. and Cook J. (1985) 'Reduced sized venting for plumbing branch lines', Heating / Piping / Air Conditioning, January.

Galowin L.S. and Kopetka P. (1982) 'Requirements for individual branch and fixture vents – recommendations for model plumbing codes', CIBW62 Conference, Water Supply and Drainage for Buildings, Lostorf, September.

Goldberg D.E. and Wylie E.B. (1983) 'Characteristics method using time-line interpolations', J. Hyd. Div. ASCE, Vol 109, No 5.

Gormley M. (2002) 'Development of a positive transient attenuator for use in building drainage systems', M.Phil. thesis, Heriot Watt University.

Gormley M. (2007a) 'Air pressure transient generation as a result of falling solids in building drainage stacks: definition, mechanisms and modelling', BSER&T, Vol 28, No. 1, pp.55–70.

Gormley M. (2007b) 'Myths and legends: developments towards modern sanitary engineering', Proceedings CIBW62 Conference Water Supply and Drainage for Buildings, Brno, September (reprinted in International Plumbing Review, 2008).

Gormley M. and Hartley C. (2009) 'From desktop to plant room: development of an innovative system for mapping and assessing trap seal vulnerabilities in building drainage systems – lessons from the field', Proceedings CIBW62 Conference Water Supply and Drainage for Buildings, Dusseldorf, September.

Gray C.A.M. (1953) 'Analysis of the dissipation of energy in waterhammer', Procs. ASCE, Vol 119, Paper 274, pp.1176–1194.

Griffiths T.J. (1962) 'The hydraulic design of drainage pipework for domestic and public buildings', Proceedings Public Works and Municipal Services Congress, London.

Harding D.A. (1966) 'A method of programming graphical surge analysis for medium speed computers', Procs. I.Mech.E., Vol 180, Part 3E, pp.88–103.

Hempel S. (2007) The Strange Case of the Broad Street Pump – John Snow and the Mystery of Cholera, University of California Press.

Henson D.A. and Fox J.A. (1974a) 'Transient flows in tunnel complexes of the type proposed for the Channel Tunnel', Procs. I.Mech.E., Vol 188, 15/74, pp.153–161.

Henson D.A. and Fox J.A. (1974b) 'Application to the channel tunnel of a method of calculating the transient flows in complex tunnel systems', Procs. I.Mech.E., Vol 188 15/74, pp.161–167.

Hope P. and Papworth M.U. (1980) 'Fire main failure due to rapid priming of dry lines', Procs. 3rd Int. Conf. on Pressure Surge, BHRA, Cranfield 1980, pp.381–390.

Hung C.K., Chan D.W.T., Law L.K.C., Chan E.H.W. and Wong E.S.W. (2006) 'Industrial experience and research into the causes of SARS virus transmission in a high-rise residential housing estate in Hong Kong', BSER&T, Vol 27. No 2. pp.91–102.

Hunter R.B. (1924) 'Minimum requirements for plumbing', US Department of Commerce, Plumbing Code Report.

Hunter R.B. (1940) 'Methods of estimating loads in plumbing systems', BMS 65 and BMS 79, National Bureau of Standards, Washington DC.

IAPMO (2003) 'US Uniform Plumbing UPC 1-2003-1'. An American Standard, IAPMO California.

International Code Council (2003) 'US International Plumbing Code', International Code Council, Illinois.

Jack L.B. (1997) 'An investigation of the air pressure regime within building drainage vent systems' PhD thesis, Heriot Watt University.

Jack L.B. (2000) 'Developments in the definition of fluid traction forces within building drainage vent systems', BSER&T Vol 21, No 4, pp.266–273.

Jack L.B. (2006) 'Drainage design: factors contributing to SARS transmission', Municipal Engineer No. 159 (Proc. ICE), March, pp.43–48.

Jack L.B., Cheng C. and Lu W.H. (2006) 'Numerical simulation of pressure and airflow response of building drainage ventilation systems', BSER&T, Vol 27, No 2, pp.141–152.

Jack L.B., Swaffield J.A. and Filsell S. (2004) 'Identification of potential contamination routes and associated prediction of cross flows in building drainage and ventilation systems', CIBW62 Conference, Water Supply and Drainage for Buildings, Paris, September 16–17.

Joukowsky N. (1900) Über den hydraulischen Stoss in Wasserlietungsrohren, Memoirs de l'Academie Imperiale des Sciences de St. Petersburg 1900, translated Simin. O. (1904) Waterhammer Proc. AWWA 24, pp.341–424.

Kelly D.A. (2009a) 'Reducing the risk of cross-contamination from building drainage systems using the reflected wave technique to identify depleted trap seals', Proceedings CIBW62 Conference Water Supply and Drainage for Buildings, Dusseldorf, September.

Kelly D.A. (2009b) 'Controlling the risk of cross contamination from the building drainage system using the reflected wave technique to identify depleted water trap seals', PhD thesis, Heriot Watt University.

Kelly D.A., Swaffield J.A., Jack L.B., Campbell D.P. and Gormley M. (2008a) 'Pressure transient identification of depleted trap seals: a pressure pulse technique', BSER&T, Vol 29, No. 2, pp.165–182.

Kelly D.A., Swaffield J.A., Jack L.B., Campbell D.P. and Gormley M. (2008b) 'Pressure transient identification of depleted trap seals: a sinusoidal wave technique', BSER&T, Vol 29, No. 3, pp.219–232.

Kelly D.A., Swaffield J.A., Jack L.B., Campbell D.P. and Gormley M. (2008c) 'A transient based technique to locate appliance trap seals within the building drainage system', BHR Group Surge Analysis – System design, simulation, monitoring and control, 10th Int. Conf. on Pressure Surge, Edinburgh, 14–16 May, pp.153–167.

Kiya F (1977) 'Hot water supply in Japan', Proceedings CIBW62 Conference Water Supply and Drainage for Buildings, Oslo, Norway, May 1977, Norwegian Building Research Institute.

Konen T.P. (1989) 'Water demand research: plumbing fixture requirements for buildings', Journal American Society of Plumbing Engineers Vol 42, No. 2, 23–62.

Korteweg D.J. (1878) 'Uber die Fortphlanzungsgeschindigkeit des Schalles in elastisches Rohren', Annalen der Physik und Chemie, 5 Floge, Band 5.

Lamoen J. (1947) 'Le coup de belier d'Allievi, compte tenu des pertes de charge continues', Bull. Centre de Etudes, de Recherches et d'Essais Scientifiques des Constructions du Gerrie Civil et Hydraulique Fluviale, Tome II, Doseor, Liege.

Lawson J.D., O'Neill I.C. and Graze H.R. (1963) 'Pressure surge in fire services in tall buildings', Procs. 1st Australasian Conf. on Hydraulics and Fluid Mechanics, Pergamon Press, pp.353–368.

Lillywhite M.S.T. and Wise A.F.E. (1969) 'Towards a general method for the design of drainage systems in large buildings', Journal IPHE, Vol 68, No. 4,.

Lister M. (1960) 'Numerical simulation of hyperbolic partial differential equations by the method of characteristics', in Ralston and Wilf (eds) Mathematical Methods for Digital Computers, John Wiley, New York, pp.165–179.

Machiavelli N. (1513) The Prince, translated by G. Ball, Penguin Press, 1961.

MacLeod K.D. (2000) 'Numerical modelling of air pressure transients resulting from detergent dosed annular flows within building drainage systems', PhD thesis, Heriot Watt University.

Martin C.S. (1976) 'Entrapped air in pipelines', Proc. 2nd Int. Conf. Pressure Surge, BHRA, Bedford, pp.F2–15.

Masayuki O. and Zhang Z. (2009) 'A study of a prediction method for drainage performance using a horizontal fixture drain branch with an air admittance valve', CIBW62 Conference, Water Supply and Drainage for Buildings, Dusseldorf, September 7–9.

Maxwell Standing K. (1986) 'Improvements in the application of the numerical method of characteristics to predict attenuation in unsteady partially filled pipeflow', Jour. of Research, NBS, Vol 91, No 3, May–June.

McDougall J.A. and Swaffield J.A. (2000) 'Simulation of building drainage system operation under water conservation design criteria', BSER&T, Vol 21, No. 1.

McDougall J.A. and Swaffield J.A. (2003) 'The influence of water conservation on drain sizing for building drainage systems, BSER&T,Vol 24, No. 4, pp.229–243.

Murakawa S. (1989) 'A study on the estimate of water demand in various buildings based on multiple regression analysis', Proceedings CIBW62 Conference Water Supply and Drainage for Buildings, Gavle, Sweden, Sept. 1989, Swedish Building Research Institute TN:18 E. Olson Ed.

O'Sullivan E.F. (1974) 'Laboratory measurements of pressure drop factors for branch discharges in vertical drainage stacks', BRE Note N3/74.

Patnaik, V. and Perez-Blanco, H. (1996) 'Roll waves in falling films: an approximate treatment of the velocity field', Int. J. Heat and Fluid Flow, pp.63–70.

Pickford J. (1969) Analysis of Water Surge, Gordon and Breach Science Publishers, New York.

Pink B.J. (1973a) 'A study of water flow in vertical stacks by means of a probe', Building Research Establishment Current Paper, 36/73.

Pink B.J. (1973b) 'A study of stack length on the air flow in drainage stacks' Building Research Establishment Current Paper, 38/73.

Putnam, J.P. (1911) Plumbing and Household Sanitation, Doubleday, Page & Co, Garden City, New York.

Reynolds, O. (1872) 'Sewer gas and how to keep it out of houses', in Allen, M. (2002) 'From Cesspool to Sewer: Sanitary Reform and the Rhetoric of Resistance, 1848–1880', Victorian Literature and Culture (2002), 30: pp.383–402, Cambridge University Press.

Schnyder O. (1929) 'Waterhammer in pump discharge lines', Schweizerische Bauxeitung, Vol 94, No. 22,23.

Schultz J. K. (2001) 'Wasting away in a hotel room – the whys and hows of Sovent', PME. Jan 21.

Standards Australia International (2003) 'Australian / New Zealand Standard. AS/NZS 3500.2:2003. Plumbing and drainage, Part 2: Sanitary plumbing and drainage', Standards Australia International, ISBN 0 7337 5496 1.

Stephens M., Vitkovsky J., Lambert M., Karney B. and Nixon J. (2004) 'Transient analysis to assess valve status and topology in pipe networks', Int. Conf. Pressure Surge, BHR Group, Chester, March 24–26, Vol 1, pp.2111–2224.

St Venant A.J.C.B. (1870) 'Elementary demonstration of the propagation formula for a wave in a prismatic channel', Comptes rendes des séances de l'Academie des Sciences, 71, pp.186–195.

Streeter V.L. and Lai C. (1962) 'Waterhammer analysis including fluid friction', Jour. Hyd. Div, ASCE, Vol 128, Paper No. 3502, Part 1, pp.1491–1552.

Swaffield J.A. (1970) 'A study of column separation following valve closure in a pipeline carrying aviation kerosene', Thermodynamics and Fluid Mechanics Convention, Steady and Unsteady Flows, 57, Procs. I.Mech.E.,Vol 184.

Swaffield J.A. (1972a) 'A study of the influence of air release on column separation in an aviation kerosene pipeline', Procs. I.Mech.E. Thermodynamics and Fluid Mechanics Group,Vol 186, 56/72, pp.693–703.

Swaffield J.A. (1972b) 'Column separation in an aircraft fuel system', 1st Int. Conf. Pressure Surges, Canterbury, England, September 6–8, BHRA Paper C2, pp 13–28.

Swaffield J.A. (1989) 'Air pressure transient propagation in building drainage vent systems', Proceedings CIBW62 Conference Water Supply and Drainage for Buildings, Gavle, Sweden, Sept, Swedish Building Research Institute TN:18 E. Olson Ed. pp.165–177.

Swaffield J.A. (2003) 'Inventors "R' Us", … thoughts from abroad', PMEngineer, May, Vol 9, No. 5, pp.22–26.

Swaffield J.A. (2005) 'Transient identification of defective trap seals', Building Research and Information, Vol 33, No. 3, pp.245–256.

Swaffield J.A. (2006) 'Sealed building drainage and vent systems – an application of active air pressure transient control and suppression', Building and Environment, 41 (October), pp.1435–1446.

Swaffield J.A. (2007) 'Influence of unsteady friction on trap seal depletion', CIBW62 Water Supply and Drainage Conference, Brno, September 19–21.

Swaffield J.A. and Boldy A.P. (1993) 'Pressure surge in pipe and duct systems', Avebury Technical, Gower Press, Aldershot.

Swaffield J.A. and Campbell D.P. (1992a) 'Numerical modelling of air pressure transient propagation in building drainage systems, including the influence of mechanical boundary conditions', *Building Envir.* 27, No 4, pp.455–467.

Swaffield J.A. and Campbell D.P. (1992b) 'Air pressure transient propagation in building drainage and vent systems, an application of unsteady flow analysis', Building and Environment 27, No 3, pp 357–365.

Swaffield J.A. and Campbell D.P. (1995) 'The simulation of air pressure propagation in building drainage and vent systems', Building Envir. 30, 115–127, 1995.

Swaffield J.A. and Galowin L.S. (1992) The Engineered Design of Building Drainage Systems, Ashgate, Gower.

Swaffield J.A. and Jack L.B. (1998) 'Drainage vent systems: investigation and analysis of air pressure regime', Proceedings CIBSE Series A: BSER&T Vol 19, No. 3.

Swaffield J.A. and Jack L.B. (2004) 'Simulation and analysis of airborne cross-contamination routes due to the operation of building drainage and vent systems', Building Research and Information, Vol 32, No. 6, pp.451–467.

Swaffield J.A. and McDougall J.A. (1996) 'Modelling solid transport in building drainage systems', Water Science and Technology, Vol 33, No 9, July.

Swaffield J.A. and Thancanamootoo A. (1991) 'Modelling unsteady annular downflow in vertical building drainage stacks', Building and Environment, Vol 26, No 2, pp.137–142.

Swaffield J.A. and Wright G.B. (1998a) 'Drainage ventilation for underground structures 1: Transient analysis of operation', BSER&T, Vol 19, No. 4, pp.187–194.

Swaffield J.A. and Wright G.B. (1998b) 'Drainage ventilation for underground structures 2: Simulation of transient response', BSER&T, Vol 19, No. 4, pp.195–202.

Swaffield J.A., Ballanco J. and McDougall J.A. (2002) 'Pressure surge in building utility services exacerbated by trapped or entrained air', Building Service Engineering Research and Technology, Vol 23: No. 3, pp. 179–196.

Swaffield J.A., Campbell D.P. and Gormley M. (2005a) 'Pressure transient control: Part I Criteria for transient analysis and control', Building Service Engineering, May 2005, Vol 26, pp. 99–114.

Swaffield J.A., Campbell D.P. and Gormley M. (2005b) 'Pressure transient control: Part II Simulation and design of a positive surge protection device for building drainage networks', Building Service Engineering, Aug 2005, Vol 26, pp. 195–212.

Swaffield J.A., Jack L.B. and Campbell D.P. (2004) 'Control and suppression of air pressure transients in building drainage and vent systems', Building and Environment, Vol 39, No. 7, pp.783–794.

Swaffield J.A., McDougall J.A. and Campbell D.P. (1999) 'Drainage flow and solid transport in defective building drainage networks', BSER&T, Vol 20 No. 2.

Teale T.P. (1881) Dangers to Health, 3rd edn, J. and A. Churchill, London.

Thancanamootoo A. (1991) 'Unsteady annular downflow in building vertical stacks', PhD thesis Heriot Watt University.

Thorley A.R.D. and Twyman J.W.R. (1976) 'Propagation of transient pressure waves in a sodium cooled fast reactor', 2nd Int. Conference on Pressure Surge, BHRA, London, pp.A2–1521.

Van Peetersen E. (1974) 'Pressure distribution in stacks', CIBW62 Conference, Water Supply and Drainage for Buildings, Dusseldorf, Danish Building Research Institute, Copenhagen, September 11.

Vardy A.E. (1976) 'On the use of the method of characteristics for the solution of unsteady flows in networks', 2nd Int. Conference on Pressure Surge, BHRA, London, pp.15–30.

Vardy A.E. (2008) 'Method of characteristics in quasi-steady compressible flows', Surge Analysis – System Design, Simulation, Monitoring and Control, 10th International Conference on Pressure Surges, BHR, Edinburgh, 14–16 May.

Vitkovsky J., Stephens M., Bergant A., Lambert M. and Simpson A. (2004) 'Efficient and accurate calculation of Zilke and Vardy-Brown unsteady friction in pipe transients', 9th International Conference on Pressure Surges, BHR Group, Chester UK, 24–26 March, pp.405–419.

Waring G. E. (1895) How to Drain a House, Practical Information for Householders, D. Van Nostrand, New York.

Webster C.J.D. (1972) 'An investigation of the use of water outlets in multi-storey flats', Building Services Engineer, Vol 39, No. 1, pp.215–233.

White P. (2008) 'A tall order', CIBSE Journal Technical File, July, pp.54–58 (bsjonline. co.uk).

White S. and Chang M. (2009) 'P.A.P.A.™ installation, Pak Tin , Hong Kong and O₂ Greenwich', Personal communications June–October.

WHO (2003) 'Inadequate plumbing systems likely contributed to SARS transmission', Press Release WHO/780, 26 September, World Health Organisation, Geneva, pp.1–2.

WHO (2004) 'Report on first meeting', World Health Organisation Scientific Research Advisory Committee on Severe Acute Respiratory Syndrome (SARS), Geneva, Switzerland, 20–21 October 2003. WHO/CDS/CSR/GAR/2004.16, WHO, Geneva.

Wiggert D.C. and Martin C.S. (2004) 'Dynamic response of laboratory exhaust systems', International Journal of Heating, Air-conditioning and Refrigeration Research, ASHRAE, Vol 10, No 4.

Wiggert D.C. and Martin C.S. (2008) 'Measurements and modelling of transients in laboratory exhaust ductwork', Surge Analysis – System Design, Simulation, Monitoring and Control, 10th International Conference on Pressure Surges, BHR, Edinburgh, 14–16 May.

William-Louis M.J.P. and Tournier C. (1998) 'Non-homentropic flow generated by trains in tunnels with side branches', Int. Jour. of Numerical Methods for Heat and Fluid Flow, Vol 8, Issue 3, pp.183–198.

Wise A.F.E. (1952) 'One pipe plumbing – some recent experiments at the Building Research Station', Journal IPHE, London, No 51.

Wise A.F.E. (1957) 'Aerodynamics studies to aid stack design', Journal Institution of Public Health Engineers, Vol 56, No. 1, pp.48–64.

Wise A.F.E. (1986) Water, Sanitary and Waste Services for Buildings, Mitchell's Professional Library, 3rd edn.

Wise A.F.E. and Croft J. (1954) 'Investigation of single stack drainage for multi-storey flats', Journal of Royal Sanitary Institute, Vol 74, No 9, pp.797–826.

Wise A.F.E. and Swaffield J.A. (2002) Water, Sanitary and Waste Services for Buildings, 5th edn, Butterworth Heinemann, London.

Wong L.T. and Mui K.W. (2009) 'Drainage demands of domestic washrooms in Hong Kong', Building Service Engineering, Vol 30, May, pp. 121–133.

Wood F.M. (1970) 'History of waterhammer', Res. Rep. No 65, Dept. Civil Eng., Queens University, Kingston, Ontario.

Woodhead C.A., Fox J.A. and Vardy A.E. (1976) 'Analysis of water curtains in transient gas flows in ducts', 2nd International Conference on Pressure Surges, London, 22–24 September.

Wright G.B. (1997) 'Mathematical modelling of sub-surface building drainage networks and their associated mechanical vent systems', PhD thesis, Heriot Watt University.

Wylie E.B. and Streeter V.L. (1978). Fluid Transients in Systems. Prentice Hall, New Jersey.

Wyly R.S. (1964) 'Investigation of hydraulics of horizontal drains in plumbing systems', Monograph 86, National Bureau of Standards, Washington DC.

Wyly R.S. and Eaton H.N. (1961) 'Capacity of stacks in sanitary drainage systems for buildings', Monograph 31, National Bureau of Standards, Washington DC.

Wyly R.S. and Galowin L.S. (1985) 'Criteria and design guidelines for reduced size vents in one and two storey housing units', ASPE Journal of Engineered Plumbing, Vol 1, No. 2, July, pp.97–122.

Wyly R.S. and Sherlin G.C. (1972) 'Performance of single stack DWV system utilising low angle stack-branch confluence and bottom shunt venting', NBS US Department of Commerce, Building Science Series, No. 41.

Yu I.T.C., Li Y., Wong T.W., Tam W., Chan A.T., Lee J.H.W., Leung D.Y.C. and Ho T. (2004) 'Evidence of airborne transmission of the severe acute respiratory syndrome virus', N. Eng. J. Med. 350, pp.1731–1739.

Zilke W. (1968) 'Frequency dependent friction in transient pipe flow', Journal of Basic Engineering, Trans. ASME, Vol 90, No. 1, pp.109–115.

Author and organisation index

Topic index